Women Who Lead In Technology

Presented by

Dr. Sharon H. Porter

Also By
Dr. Sharon H. Porter

www.drsharonwrites.com

Next In Line To Lead : The Voice of the Assistant Principal

Class of 2017! What's Next?

Women Who Lead: Extraordinary Women With Extraordinary Achievements Volume 1

Women Who Lead : Extraordinary Achievements With Extraordinary Achievements Volume 2 Featuring Latina Leaders

Women Who Lead in Education Volume 3 Featuring School Principals 1st Edition and 2nd Edition

Young Ladies Who L.E.A. D.

The Power of Networking: How to Achieve Success With Business Networking

North Carolina Girls Living In A Maryland World

Fifty & Fabulous The 2019 Edition

The HBCU Experience Anthology Volume 1 The North Carolina A&T State University Edition

The HBCU Experience Anthology Volume 2 Alumni Stories From the Hill of Kentucky State University

Dr. Tracy Daniel-Hardy

About Dr. Tracy Daniel-Hardy

Dr. Tracy Daniel-Hardy is the Director of Technology for a public school district in Mississippi. Dr. Hardy has over 25 years of experience as an educator in various capacities including teacher, guidance counselor, and instructional technology specialist. Dr. Hardy is a Tech & Learning Advisor, Future Ready Tech Leader Advisor, Immediate Past President of Mississippi Educational Computing Association, the Mississippi Department of Education's Technology Advisory Committee, Mississippi Educational Technology Leaders Association, Delta Sigma Theta Sorority, Inc., 2021 Board Member of Lighthouse Business and Professional Women, and many other civic and community organizations. Dr. Hardy was the recipient of the 2019 Success Women's Conference Volunteer of the Year Award, 2018 PowHER Education Award, 2017 NAACP Educator Award, 2016 Alcorn State University National Alumni Association Professional Achievement Award, the Lighthouse Business and Professional Women 2014 Woman of the Year Award, and the Blessed Gyrls Rock Award. Dr. Hardy champions causes and movements within her affiliated organizations that promote and encourage equity specifically for children, women, people of color, and others who are disenfranchised, under represented, and marginalized.

She is also co-author of **Shut'em Down: Black Women, Racism, and Corporate America** Anthology and has been featured in Tech & Learning magazine and Gulf Coast Woman magazine.

Foreword

It is February 2021 as I write this. The United States has just finalized the most contested and tumultuous presidential election that I have observed in my nearly 50 years of life. This election was fueled by a divisive political climate due to racial unrest. (Or was the racial unrest caused by the divisive political climate?) We have recently witnessed an insurrection of our State Capitol Building and have also witnessed women breaking political barriers. According to the Center for American Women and Politics, a record-breaking total of 144 women were serving in the United States Congress last month. Additionally, 52 of those women were women of color (Black, Indigenous, Latina, Asian, and Pacific Islander). We saw the election of the first woman vice president and witnessed women like Stacey Abrams (from my hometown) fight like hell to change the composition of our United States Congress. Women everywhere are feeling excited, inspired, empowered, and simultaneously annoyed, ashamed, and afraid because of the recent events. We are in a crucial time in our history and women are on the front line.

We are, also, still in the middle of a pandemic. The COVID-19 virus has ravaged the United States with over 28 million reported cases and 500,000+ reported deaths to date according to the CDC's COVID Data Tracker. However, there is now a glimmer of hope. Coronavirus vaccines are being distributed and made available to Americans. It seems as if we may be turning a corner as the numbers of positive COVID-19 cases have started to drastically decrease in many states.

I cannot say with certainty that the decline in new positive cases should be attributed to the increase in those who are being vaccinated, but the timing of the two events aligns. The downward trend of the virus gives us hope of returning to a normal life of working, socializing, and traveling. Therefore, we must acknowledge some incredible women who have worked in STEM (science, technology, engineering, and mathematics) contributing to the coronavirus vaccines and the possible commencement of the eradication of the COVID-19 virus.

One of the women, Dr. Kizzie Corbett, an African American scientist with a PhD in microbiology and immunology, worked on the team that helped to develop Moderna's messenger Ribonucleic Acid (mRNA) vaccine. Dr. Corbett's contribution to this vaccine that has a 94% to 95% efficacy rate is more than just about eradicating a virus that caused a pandemic, it is also about representation. Dr. Corbett understands the importance of speaking out and being heard to encourage those who are underrepresented like young people, people of color, and women.

Like many, I was hesitant about getting the coronavirus vaccine because it seemed to have been developed too quickly until I read several stories about Dr. Katalin Kariko and her 30 years of research on the mRNA. Dr. Kariko was demoted, had her work constantly rejected and was even considered a quack by some due to her perseverance to the research of the mRNA. However, it was the diligent work of Dr. Kariko that developed the mRNA research for the Moderna and Pfizer vaccines. Her groundbreaking contributions have yet to be felt, seen, or fully understood in my opinion.

Since the beginning of time, the woman has been the center of the home. The woman has been the glue that kept everything together in the home as she cooked, cleaned, bandaged, washed, mended clothing and hearts, disciplined, encouraged, and maintained order. Although the woman played such a significant role in the home, she was thought to have no power or authority in the home. She was undervalued like Dr. Katalin Kariko was in her field. The power and authority were placed in the hands of the man of the house. He made the decisions, the money, and had the last word. But let us be truthful about it. In many instances, the man of the house was just the figure head while the woman of the house actually had the power, the influence, and the plan.

Much like the hidden figures depicted in the movie with the same name, women in technology and their roles can be likened to the woman of the traditional home. Her presence was needed and noticed, but underestimated. She often gently fostered ideas, while encouraging others and maintaining order until someone noticed she had something profound to say and decided to listen to her. Even then, her contributions were minimized and either excluded or briefly mentioned. However, we cannot adequately or fully discuss the history of the technology industry, or the history of anything, without mentioning the roles of groundbreaking and trailblazing women. Oftentimes, these amazing women did not set out to blaze trails, but instead tried to do what women have traditionally done - create solutions for problems.

Although the typically male authors of history exclude women from the contributions to the technology field, there would be little to talk about without acknowledging the contributions of women. Journalist Claire L.

Evans said it best when she said it takes "more than dudes in a garage to imagine and build the future". It takes the hard work, grit, and talent of women like Dr. Nashlie Sephus, the Applied Science Manager for Amazon's Artificial Intelligence (AI) team to help build the future of technology. Dr. Nashlie recently purchased 12 acres of land in her hometown of Jackson, Mississippi to do just that, launch a technology hub that will expand what she voluntarily does weekly to grow technology access and skills of the Jackson, Mississippi community.

Without women, the world would be void of many of the conveniences, innovations, and luxuries we enjoy today. As a woman educational technology leader and a woman of color, I have experienced many obstacles and challenges while trying to create opportunities for conveniences and innovations. I am now inspired more than ever by the movement in our country to build seats where none exist.

I now present this edition of Women Who Lead in Technology.

-Dr. Tracy Daniel-Hardy

Contents:

ACKNOWLEDGMENTS

I would like to acknowledge the contributing authors of Women Who Lead In Technology.

There continue to be conversations about gender diversity in the technology industry, but women are still underrepresented. According to a report from TrustRadius 72 percent of women tech report being outnumbered by men in business meetings by a ratio of at least 2:1. Twenty-six percent report being outnumbered by 5:1 or more.

Thank you for sharing your journey.

Foreword: Dr. Tracy Daniel- Hardy

Afterword: Dr. Sarah Thomas

Contributing Authors

Alexis Nicole White
Dr. Carol Gorst
Chanel Johnson
Dr. Jocelyn McDonald
Karen Walsh
Kenyatta Powers-Rucker
Laura Hart
Mandy Froehlich
Patricia Brown
Dr. Sonja Jones

Alexis Nicole White

"Surround yourself with the subject matter experts"

-A. White

Chapter 1

"If I drive my career, where is the vehicle?"

By Alexis Nicole White, PMP, SMC

During *The Great Recession of 2008*, I watched my dreams of becoming an entertainment and news journalist shrink faster than most people's 401K. Greeted by a world of instability and uncertainty, I was entering a world of economic chaos. My senior year was less than comforting or exciting. Enron collapsed right under the nose of its employees. Then, Harry Markopoulos exposed Bernie Madoff as a fraud with his Ponzi Scheme. Simultaneously, the housing market crashed as millions of people faced foreclosure and eviction. Then, a significant pipeline for gasoline had broken during the hurricane season, and gas was roughly $5 a gallon. Most of America was living in crisis mode, and it was indeed survival of the fittest.

As with most millennials, I lived in my bubble during this time. Infatuated and consumed with relocating to Atlanta, I was naïve and unaware of the 'big bad real world' that I was walking into blindly. Like most students during this time, I graduated jobless. Yet, I took a bold leap of faith and moved to Atlanta, anyways, with $150.00 to my name, hope, prayer, and desperation. And, by the time I arrived, I only had $50 left. (I was young and full of spirit, as the Elders would say.)

Initially, I was unapologetic about being both unemployed and broke. In the beginning, it was cute and fun. Since I was jobless, I had complete flexibility over my schedule to attend all of the latest industry events. Although my hobbies in entertainment were vibrant, they were not paying. I was in the loop of all of the who's who in Atlanta. Yet, when reality set in that going from couch to couch living from pillar to post was not going to cut it, I decided to use my networking skills to seek employment.

After undergoing plenty of interviews for three months, nothing materialized. Most of the jobs were less than ideal for a "recent college graduate". I subscribed to the notion that I would graduate and make plenty of money. I expected job offers to roll in with the wind. Nonetheless, our generation started to combat the whole "college is the way to go" expectation that our parents set for us because we did everything right, and still, we had nothing. While some graduating were dealing with the reality of having false expectations set by people who college did work for previously. Now, graduates like myself were competing with the seasoned vets terminated from their previous employment, and they needed work to pay their bills also.

It was exhausting to undergo multiple rounds of interviews for jobs paying $8 - $12 an hour. However, I needed to do whatever I needed to do to foster my independence. Rather than promoting my entertainment work and featuring up-and-coming music artists on MySpace and Twitter, I decided to post a news alert for "Who's hiring?" and included a vulnerable post asking for help. Immediately, I received more dead-end job leads. However, one from a person who worked at Dish Network invited me to a job fair. Although he admitted that he could not promise anything, I decided to show up anyways.

I barely had enough gas in my car to get from Lenox Road in Buckhead to northeast Marietta. I borrowed $10 to get me there and back. Moreover, an unpaid speeding ticket resulted in my license suspension, and my insurance was invalid for my car. My shoes started to "talk," and my synthetic wig turned "nappy" at the ends. My Tahari suit barely fit, but I stuffed my butt in my slacks and went on about my day. I showed up anyway. Even though I felt deflated and defeated, showing up was one of the best decisions that I ever made.

Though this would also be another lower-paying wage job, it exposed me to technology.

Whereas the role was less than fulfilling financially or intellectually, it taught me how to be present in technological space. I learned how to run cables, install devices, and leave the rooms presentable. Although I can still hear the phones ringing off the hook in my head now, I learned the value of interpreting and communicating technical concepts, identifying ways of improvement, and implementing strategic goals to motivate my team to enhance our products' delivery. I learned to collaborate with the front-line technical teams and senior leadership, from general managers to operations managers and regional managers. Although I met all of my performance goals and grew to understand the value of "knowing your business," I was not unmet with adversity. I was accused of sexually inappropriate behavior when my territory was the worst in the southeast region to be one of the top areas within six months. Selected as "Employee of the month," I decided to resign. Making $23,920 a year was unacceptable. Neither were their apologies acceptable for insulting my character empowered me to want to stay. I felt slandered, defamed, and disrespected, which was not coming at the cost of $11.50 per hour. Finally, although their accusations were unfounded, it did not stop other people from assuming the rumors were true.

Nonetheless, because of that one job at Dish Network, I developed the rare skill of dispatching. By the following summer, a recruiter called to support Ericsson within their field service operations division. Ericsson undertook a five-year contract to upgrade, maintain and correct any of Sprint's cellular towers. At this time, we were implementing all of the new 4G technology, decommissioning the old IDEN (2G/analog) sites, and expanding the network for Sprint. Additionally, there were several other initiatives that I was able to support within the three years of service with them. Although I grew my income from $23,920 to $55,000 a year, I was looking to be more progressive.

As I started to interface with the new buzzword "project manager" or "project coordinator," I perceived that I could easily do their job. It didn't seem like rocket science, more like a lot of administrative coordination. Additionally, I seemed to know more about the technical installation than they did. Once I started to meet real-life project managers who were making over $100,000 a year and were living the good life, I wanted in on this lucrative field. Although it may sound superficial, I wanted to increase my income as well as gain stability. I wanted to grow and thrive, work fewer hours and make more money. I had enough of all of the 18-hour days,   s and I ran thin. I had become so accustomed to the overtime that I couldn't afford to live without it. However, after realizing that I couldn't recall what I exactly did with three years of my life (aside from working), I decided that I would no longer be "overly" available to corporate America. Secondly, the window for growth closed at Ericsson. I was clear about getting into project management, but there was no effort from my leadership team to help me navigate that pathway while at Ericsson. Yet, again, as "Employee of the month," I was resigning for a better opportunity.

Whereas many people still could not fathom why I would resign from a phenomenal company such as Ericsson, I understood the concept that they were continuously preaching. "You drive your career." It went to my head! After repeatedly undergoing phenomenal performance reviews, making 5/5 year over year, I didn't see a path forward. Although I secured every pay increase and bonus, thoroughly enjoyed my nine and a half weeks of paid time off every year with 100% of my benefits being paid for by my employer, it was not enough for me to show up and to be satisfied. I was serious about growing my career. I demanded a work-life balance. I desired to be able to make more money with less effort. I wanted to continue to grow with current technology, and I wanted to be compensated for my leadership efforts, none of which was happening for me at Ericsson at that time. Subsequently, I learned that if you're too good at your job, it's comfortable for management to leave you where you are because the problems are being solved with your efforts.

I championed change for our department. I coded, tested, and approved a new interface to help create efficiencies among the team. I even trained over 100+ people on the latest software, wrote the processes, served as a mentor for new hires and beyond. Never was I ever promoted voluntarily to another position. However, several "underperforming" colleagues were identified as the latest leads in our department; and they did less work than I did. Witnessing things like that put a bad taste in my mouth because it forced me to comprehend "you drive your career." Although they gave me the vessel to navigate, they did not give me that car. Therefore, how was I to drive my career? The only way I could do so was to look for opportunities beyond the organization.

Repeatedly, it was the previous skills that I mastered that opened up doors for me that no man could shut. I became more technical. As I was involved in the tactical, day-in, and day-out minutia of the work, I absorbed as much as possible to "know my business." Because I knew how things work, I could offer a strategic approach to keeping our customers happy but delivering products that were not only helpful but useful. I became overwhelmed with the need to create a pleasurable customer experience. I never wanted anyone to ever call me and feel underserved or misled.

As I've continued to grow my professional career, I have facilitated many different essential initiatives, starting as a dispatcher to now being a certified project ger and scrum master. Still living by the motto of "you drive your career," I've led multi-million-dollar projects for some of the world's top companies in both the private and public sectors. From supporting a major LTE upgrade at AT&T to revamping the entire Indiana State Department of Health's telecommunications program, I've singlehandedly led a lot of technologically advanced projects.

From mobile technology to wireline, Voice over IP, to audiovisual augmentations down to basic infrastructure, I've expanded my reach from basic at-home installations to customer-premise equipment. I've mastered facilitating conversations from C-level executives down to our technical support teams. I've collaborated with engineers, designers, architects, application developers, and business analysts. I've partnered with information security teams and other stakeholders to ensure the alignment of our project's solutions and provided best practices based on my expertise from leading such projects.

However, I would have never reached this level of success had I not been open and willing to learn. I probably worried enough people raggedy researching unfamiliar topics, asking questions, and being persistent about getting a suitable answer. Sure, I've been discriminated against for being a woman of color. I've dealt with some of the most irritated, cranky, and grumpy folks. Many did not like me, my direct or blunt approach, and I've been labeled many different negative things. I've been overlooked for promotions, not because I was unskilled or underqualified. I failed to be promoted because I've lacked "soft" skills. I was told I'd hurt a grown, six-foot-three man's feelings because I asked clarifying questions regarding a project.

Nevertheless, I learned to put myself in their shoes and be flexible with the inflexible. As a woman in corporate America, I can't afford to be soft, passive, or weak. I crack jokes. I offer lunch. I try to get my team warm up to me and be willing to support them when they're in need because they will have my back when I need them. Unfortunately, I quickly realized that I could not expect upper management to support me nor my decisions. Many times, they agree to do the exact opposite of the proposed recommendation. Yet, I'm expected to fix their missteps even though they go against better judgment.

Even though I have been left disappointed by the decisions made by senior leadership, I am often praised for my abilities to get the job done. Not only have I been requested to support complicated projects but, I've also saved several failing projects within my first 90 days of joining the organization by simply listening to the customer and responding to their concerns. Moreover, I can mitigate risks by asking critical questions and exploring avenues not yet traveled.

I wish that I could say that my journey has been effortless. It's been a lot of aggravating and frustrating moments. Without using my race, sex, or gender as an excuse for some of the professional hardships I've experienced, it is evident that there are times when I have a seat at the table, and I represent diversity. I refuse to admit that even in 2021, I am expected to simulate into organizational cultures to make senior leadership more comfortable. I've experienced jokes about my ever-evolving hair, comments about my car and handbags. I've been criticized for how I communicate with my teams. Because I'm generally laughing, smiling and cracking jokes, I am misunderstood as being 'too flirtatious' or 'dingy'. If I am direct, straight to the point, I am mean and off-putting. The list goes on and on... and all I can wonder is "if this were a man-to-man conversation, would this be a real issue?"

Subsequently, I've learned how to be less emotional but emotionally intelligent. I've learned how to read my teams and respond accordingly. Additionally, I've learned how to tailor my approach and delivery depending on the audience. Sadly, the reality is that in some environments, if you do not look like and speak like those in senior leadership, you're unable to thrive. For that, I encourage women to explore organizations that are truly aligned with their values and beliefs. We drive our own career. If you're working and feel underpaid, do not hesitate to find a company that will properly compensate you. If you feel as though your voice is not being heard by management, discover an organization that honors and open-door policy and values the feedback from its employees. If you find yourself being one of the top performers, but you're unable to break the glass ceiling, I encourage you to identify employers who are looking for someone just like you. Believe in yourself. Take risks. Learn from poor choices and bad decisions; yet, empower yourself to grow, thrive, and be successful. Do not be loyal to a company that does not support you.

For me, my desire to grow was always driven by monetary gain. My goals were always surrounded by me making X number of dollars an hour, and so forth. For you, it may look like something totally different.

Technology is forever evolving and growing. I encourage you to know that you can do this. It does not have to be a formal or traditional learning path, either. There are several methods that one can use to be leaders in a growing field. Nonetheless, you can do this. Although your path may not look exactly like mine, here are ten key takeaways from my journey that can help you excel in the technology field, without a formal degree or certifications.

Alexis Nicole White's Technology Tips

1. **Never be afraid to ask questions, even the stupid ones**– no question is ever dumb too. Often, you raise awareness to mission-critical issues that could be deemed a 'show-stopper' or inquire about a topic that no one else considered.

2. **Don't make assumptions about anything** – nothing is as straightforward as it seems.

3. **Document every conversation utilizing OneNote** (meeting notes, project notes, m notes, etc.) – retaining information (documentation) can help revitalize one's memory and cover you from missteps, especially if you're taking action on an item from a senior leader in your organization.

4. **Learn as much as you can when it comes to relevant technologies** – never become complacent with what you know; things are constantly changing..

5. **Understand how to read the drawings** (wire drawings, schematics, rack elevations, and the room/site architecture drawings) – understanding the drawings will enable you to have technical conversations with the team. It guides your understanding, understand what technologies are plugged in where, and identify where data drops are going, cables are run, electrical outlets are,

arrangements of the furniture, and other critical information for deployments.

6. **Read and understand the scope of work inside and out** (and make notes) – the statement of work tells you what's included, what's not included, the assumptions, constraints, and the high-level risk.

7. **Ask more questions and document the outcome of those conversations** – cover yourself from anything backfiring on you.

8. **Surround yourself with the subject matter experts (SMEs) and learn all you can from the**– professional development, and ongoing growth is best for elevation in your career.

9. **Participate in every conversation and be comfortable with the unknown** – By surrounding yourself with people who can have formal and informal discussions that will help your understanding. Speak up, ask questions, and forge relationships with people to feel comfortable with sharing information with you that you will need to know to optimize your success.

10. **Conduct independent research on related topics** – Expand your knowledge base continuously.

About Alexis Nicole White

Alexis Nicole White is a senior project manager and scrum master with over ten years of experience supporting enterprise Information Technology and Telecommunications Services Senior Project Manager specializing in the digital supply chain, mobile workforce, wireless, VoIP, audiovisual, and infrastructure augmentations.

Find her on LinkedIn.com at
https://www.LinkedIn.com/in/alexisnicolewhite.

Dr. Carol M. Gorst

"Be excited, not afraid of tough projects"

Dr Carol Gorst

Chapter 2

Always Be Learning

By Dr. Carol M. Gorst

I grew up in rural Wisconsin on a farm. I was the fifth of seven kids and our family lived miles away from the closest small town of 700 people. Growing up on a farm meant doing chores first and everything else second. It was the literal, "make hay when the sun shines" motto. When the hay was ready; you baled it. When the fields were dry enough; you planted crops. You quickly learned that circumstances other than 'what you wanted' dictated schedules.

Everybody had a role to play in making the farm run smoothly. Even at age five, I had my chores. It was assumed you knew when to do your chores, and it was expected they were completed without anyone checking on you. Consequences were never good if you forgot. Forgetting meant some essential activity did not get finished. On a farm, that was never a good thing. I definitely learned responsibility at a young age.

I was one of the few kids who cheered for the first day of school. I could not wait to have school as a reason to not be home doing chores. For some reason, during the school year, doing homework was an acceptable reason to get out of doing chores. I brought home lots of books every day to make sure I had a steady supply of homework. Backpacks were not a thing when I was a kid, so I carried home a big stack of books every day. I could hardly see over the top of the stack. I always brought home more books than I needed. I was not trying to inflate the apparent size of my homework; rather I simply could never decide what subject I wanted to work on each day.

When there was time off from chores, the options for fun were limited. Entertainment was usually playing games with brothers and sisters or entertaining myself. With seven kids, there were frequent disagreements about the "rules" of games. My Mom found a way to separate us for a "time out" when that happened. For some reason we had a bookcase full of Encyclopaedia Britannica and the Britannica Book of the Year starting around 1961 and going forward. I guess those were the Wikipedia of my childhood days. I did my best to have my time outs assigned to the chair in the corner by the bookcase. I can remember spending hours flipping through the books looking at the pictures and reading about places and people that were very different from my life in rural Wisconsin.

I was an inquisitive child. No matter what we were doing, something always made me curious. Reading the encyclopedia all the time generated more questions. I bombarded my parents with questions about topics irrelevant to daily life on the farm, but that were very important to me in the moment. I guess all of my questions prompted my Mom to buy me a science kit with a microscope. I remember spending hours searching for things that I could put on a slide and inspect. It is possible that entertaining myself for hours in my microscope quests provided my Mother some much-needed respite from my constant questions.

My Mom worked all day on the farm helping to tend the fields as well as growing food in the garden and canning food for the winter. When her workday at home finished, she prepared the food for the family and went into town for her job as a cook in a local diner. One of my chores was to reheat supper for everybody. For those of you uninitiated, lunch is known as dinner, and dinner is known as supper on the farm. It's a country thing.

When my Mom returned home from work, I can remember her sitting me down at the kitchen table and asking if I had finished my homework. She would quiz me on my vocabulary or make me do math flash cards. For as much as I supposedly loved to do homework, I resisted these quiz sessions. It was always the same exchange, "Why do I have to do this?" I would ask. She would reply, "You need to get an education so you don't have to depend on a man when you are older." It was a different era when she was finishing high school. Girls got married and had families. Sadly, I would discover in my own high school years that those attitudes had changed little through the years in my small town.

'Get an education' evolved into 'always be learning' over the years. I cannot count the amount of times that advice has served me well.

Anytime I felt stuck in a job, I would think about what I could learn to improve my chances for a new or different role within my company. In the fast-paced, constantly changing world of technology, I cannot think of any better advice than "always be learning."

As much as I loved going to school, and as strongly as I believed that an education was important, I lacked knowing how to get an education beyond high school. Fortunately, a couple of teachers came into my life that helped lead the way. Mrs. Kumm, one of my grade school teachers, never let me get away with taking an easy assignment. If she suspected I wanted to take the easy task she would say, "You are capable of more; you must challenge yourself. How else will you succeed when you go to college?" Mr. Jarocki, my seventh grade science teacher, let me do extra projects to learn things that were not part of the normal curriculum. When he discussed the projects, he always did so in the context of how it would help me in college.

These teachers had planted a seed. My first goal, therefore, would be to go to college. I still had no clue how to do that, however, I knew it would cost money. Therefore, while I figured out the logistics of how to go to college, I decided I was going to have to earn a lot of money. Somewhere in my seventh grade year, I remember announcing to my Mom over breakfast that I was going to college and I was going to need a job to help pay for it. I remember her saying, "You are? Well, I think Dad is expecting you to help around here this summer." As the summer break from school approached, I announced to my Father that I needed to get a real job to earn money for college. "You have a job, here on the farm," he replied. I said, "Then you are going to have to pay me so I can earn the money!" While he did not seem to initially like the idea, he also did not seem surprised by the request. In retrospect, I am sure my Mom had prepared him. Besides, my Father had to hire a couple of people to help with baling hay and other farm work because my older siblings had left home. After several negotiations, we agreed that he would pay me for doing "farm work" which would be any work he paid someone else to do. He was not; however, paying me for doing my regular chores. On my first day of work that summer, I asked the hired hands the amount they were paid. Their hourly wage was the same as mine. That seemed right to me. It never occurred to my thirteen-year-old self that they were grown men and that they were doing a higher percentage of the work than I was. I guess my Dad believed in equal work (effort), equal pay. I assumed that is how it would be in the future when I had other jobs. That assumption was wrong and I would learn that lesson many years later.

For a few years, I was working my plan. I was saving money for college and I would go to college when I finished high school. As high school progressed and my senior year approached, it dawned on me I had no idea how to apply for college. I had an older sister that had gone into a nursing program through the army. I did not know anyone who had applied to the University. After a disappointing conversation with the guidance counselor, who explained to me there were good jobs to be had "around here" and there was no reason to "worry myself" about college, I angrily went off to chemistry class in a very bad mood. I was fortunate my teacher noticed something was wrong and asked me about it. I explained the guidance counselor said he was not giving me the college application and I did not know how to get one. Mr. Schultz, my chemistry teacher, muttered something that did not sound very nice, and stormed out. He came back several minutes later with an application. He handed it to me, instructed me to fill it out and said I could go to any school I wanted. Later that year, Mr. Schultz informed me and several of my classmates that we were driving to the extension college in the nearby town to take a college entrance exam in a few weeks.

The seventies produced Star Wars and the first desktop computers. However, computers were far from common in schools. Mr. Schultz was also the person who introduced me to computers. It was not even a computer when he first shared it with my class. Rather, it was a build your own computer kit. Enthusiasm for assembling the pieces escaped me. However, the stories of what computers might be was very interesting. When presented with the chance to go to another school for a day to work on word processing systems, several of my classmates and I signed up for the challenge. It opened a window into the types of applications that computers could offer. I credit Mr. Schultz for preparing me for many of the opportunities that have come my way since I left my hometown.

College was an entirely different world. I may have learned responsibility at a young age but that wasn't enough to get me through my first year of college. I had no clue what to do. How to navigate the campus, finding classrooms, and even finding the correct cafeteria all felt a bit overwhelming. My high school graduating class was less than 70 students.

There were more than 70 women on one floor of my dormitory. Worse, I was not sure what major I wanted! The information to help with class selections broke out by majors. I had no idea what classes to take. The most common advice I received when discussing ideas for my major was "do something you love". I had no idea what I loved. I was a college freshman. I joked with an advisor that if there was a major in beer and pizza I was ready. The best advice I received after acknowledging that I had no idea what type of career I would love was "you will figure it out eventually". After years of working in science and technology, I can safely say it is okay, as you begin your college career, if you don't know what you love. It is equally important to understand what you do not like. The best you can do is engage in projects or take classes in things you find interesting. If the fit is not the right one, that fact will always show itself. In the fast-paced technology world, careers exist today that couldn't be conceived of ten years ago. Do not stress about what you should do. Make the best of your current situation and honestly look at what is working and what doesn't. Look at the opportunities that fit into your area of ability or your interests and question how you can move forward or change course.

At the beginning of college I knew I liked chemistry and I had thought about becoming a veterinarian. I had taken some of the basic science classes for pre-veterinary medicine along with psychology and several other freshman-level required classes. I tried several different majors and took a wide variety of classes my first two years at University. At the beginning of my junior year, I knew I wanted to graduate in four years so I took the reverse approach. I reviewed all the courses I had completed to decide what majors I could choose and still graduate in four years. To my surprise, there were only two options. One of them was chemistry. So that day moving forward, I became a committed chemistry major.

The chemistry department at the University of Wisconsin La Crosse required students to meet with a faculty advisor their junior year to make sure they chose the proper classes to satisfy their major. The professor advising me said I would be required to take organic chemistry, physics with a calculus prerequisite and calculus my next two semesters. I had dropped calculus my freshman year of college, so it surprised me I did not walk away and just pick another major to extend my college experience. Having to pay for college on grants, part-time work, and student loans was a good incentive to keep me focused. I said, "Sign me up!" Surprisingly, the professor did not say this is impossible or not realistic. He did say this is a very hard schedule and asked if I was sure I wanted to do it. I was sure. He signed me up and made sure I knew where the tutoring office was for chemistry. He recommended that I take no other classes, except a PE class, to round out my credit load for the semester. I am happy I took his advice on the PE class because it was one of the hardest years of my life. While true at that time, it would turn out as one of the best opportunities to prepare myself for some of the years that were to come.

The next two years proved challenging, interesting and again influenced by some wonderful teachers. Dr. Roland Roskos, physical chemistry, and Dr. Jerry Rausch, organic chemistry, were the two most consequential professors and they would greatly influence the path I would follow in the years ahead. Dr. Roskos would challenge me with one of my first projects using computing in chemistry. Dr. Rausch was a wonderful educator and motivator. As I progressed through my classes, I once again faced the dilemma of what I wanted to do. I was a chemistry major, but what was I planning to do after graduation? With the aid of both professors, I obtained a summer internship at the Department of Energy Ames Laboratory at Iowa State University. I was unaware of internships and did not appreciate how important it was to have real-world experience.

After graduation, I decided to go to graduate school for my Ph.D. I based the decision partly on it being a natural progression and partly on being geographically limited in my options. At the time, I was married to a medical student in Milwaukee, WI. He was not open to my living elsewhere. The chemistry jobs in the area put limits on my opportunities. I had enjoyed the research experience during the internship at Ames Laboratory. I earned a modest salary as a graduate student that was enough to cover essential expenses. Graduate school was challenging in multiple ways. Time management, learning new and difficult topics, teaching chemistry to undergraduates, executing my research projects and finally writing a thesis were new experiences. My research involved data collection using a variety of instrumentation. By the time I entered graduate school, computers became more common and I spent much of my time on a console collecting data or on a computer processing data.

As my time in graduate school ended, so did my marriage. However, the doors closing in graduate school and my personal life gave me the opportunity to explore every option that presented itself. Dr. Stephen Ragsdale, my thesis advisor, had opened many doors and provided many opportunities to network and grow personally and scientifically. My first trip abroad was for a scientific conference in London. It was a great experience and an opportunity to discuss career ideas with a diverse group of scientists with a broad range of life experiences. I encourage anyone pursuing a graduate degree to take time away from their research to attend conferences and develop their professional network. I decided that I would continue my studies in spectroscopy. I enjoyed the challenges of collecting the data and the ever-increasing variety of software applications required to process information or to crunch the data through various algorithms. I had secured interviews at two institutions that could not have been more different. Both Professors extended Postdoctoral offers. One offer was from Johns Hopkins University and another from the University of California, Davis (UC Davis). Every fiber in my being said to take the opportunity from Johns Hopkins but my inner compass was pointing to UC Davis. Shortly after the interview at UC Davis, the Johns Hopkins professor called insisting I decide that day. He had other people to whom he could offer the position. The further you progress in your career, the more likely you will meet someone who will try to bully you towards something that doesn't feel right. I have learned to listen to my inner compass through the years and urge you to do the same.

I interrupted my postdoctoral research at UC Davis with one year as a visiting scholar at the E.C. Slater Institute, University of Amsterdam. Living abroad gave me confidence that I could rely on myself to take on bigger challenges. While in Amsterdam, I decided I would start applying for Assistant Professor positions. During the various research positions I had learned a multitude of techniques. I was comfortable moving between different spectroscopy platforms and computer operating systems. I felt ready to start my research program. As I began applying for positions, I realized that I was consistently the university's second choice when I had been a candidate of interest. It occurred to me that I might be holding myself back. Did I really seek being a professor or was I applying for these positions because it was the only next step that I knew? I reached out to a former colleague from UC Davis. He was working at a software company specializing in scientific research software applications headquartered in San Diego, California. I sent him a copy of my résumé and asked if he would review it. I did not know how to go about finding positions outside of academia and was seeking his advice.

A few weeks after sending my résumé, we both happened to attend the same conference. I sought out his company's exhibit and set about trying to find my friend. Instead, I found his manager. She asked if she knew me; my name seemed familiar. I explained we had not met, my colleague from UC Davis was the only person I knew at the company. She asked if I had sent him my résumé, to which I replied yes. She asked if I wouldn't mind waiting there a few minutes. She returned with some of her team members. I quickly realized that I was in the middle of an ad-hoc interview. A few weeks later I was formally interviewed in San Diego for a position supporting the company's spectroscopy software applications. I was thrilled. This path felt like the right one. During the interview, I was disappointed to learn they had just offered the San Diego position to someone else. They were still working out the start date with that candidate but she had accepted the position. A similar position was available on the east coast if I was interested. I somewhat jokingly said to the hiring manager that I could start in two weeks if they wanted to offer me the position in San Diego and give the other candidate the position on the east coast. Unfortunately, I returned home with a verbal offer for the position on the east coast.

Upon my return, one of the school's that notified me before that I was their second choice called to extend an offer. Their first choice candidate had dropped out. I was surprised to find that I was not jumping to say yes. I told the University that I was still interested but was starting discussions for another offer. The University made a good offer, but I stalled while I negotiated with the software company. The original candidate for the San Diego position was still wavering on a start date. I mentioned I had a brother in San Diego and would happily accept and relocate if they offered me the position in San Diego. During this time, the University called back with more research funds. This interested me, but I was honest that I was also negotiating another position. In the end, the San Diego company took me up on my offer for the west coast position. The university had increased their offered research startup funds three times during the interim. I learned two things from this experience. If you ardently pursue something but you continuously fall short, perhaps you are trying to fulfill someone else's expectations not your own. You may be instinctively holding yourself back because you know you are on the wrong path. Secondly, I learned to ask for what you want.

The software applications I supported in my new role ran on Solaris. I had minimal experience with the Solaris operating system. I ran the applications in my earlier roles, but never had to mess about installing or configuring the software. I spent a great deal of time my first year asking a lot of questions and writing extensive notes so I could correctly execute the installations without help. When I visited customers, I would always be partnered with their resident UNIX expert. Sometimes they were very dismissive because of my limited UNIX experience. My best was the most that I could offer. If someone wanted to dismiss me because they were smarter than me at one thing, it was their prerogative. However, I committed myself to making sure everyone I visited had their issues resolved. I promised to get answers when I did not have them. I pulled in the right people to fix the problem when I could not. These experiences lead to a big take away. Over time, most of those people respected my abilities even when my UNIX acumen fell beneath their standards. More than one has reached out with open positions through the years. Keeping your word and being the solution to a problem is powerful currency. Brand yourself as a problem solver and people will want you on their team.

I transitioned between multiple software applications and a few companies during my time in San Diego. The one constant through everything was no matter the technology, eventually it would fall short because someone would stretch the limits of the application. Technology is constantly evolving and to stay current you must keep pace. You cannot assume what is good enough today will suffice tomorrow. I left San Diego to support the regulatory affairs software applications for a biopharma company. Again, there were all new applications. New reporting tools, new scripts and new data analysis tools to learn. However, many of the challenges for the users were the same as before. The technology failed to meet expectations or the needs of the organization changed and the technology no longer sufficed.

Working in a very large company for the first time presented some new challenges. There were considerably more processes in place. Sometimes the process strangled the ability to get work completed. There were also new opportunities. In a small company you wear many hats making it difficult to fully grow into a role. Large companies have training programs and offer opportunities to learn new skills. If you are lucky enough to work in an organization that can support training and new skills development, it is essential to take advantage of it. Always increase the depth of your technical knowledge when the opportunity arises.

As satisfying as my time was at the biopharma company, I also felt something lacking. Large companies often are well-meaning, but fall short in supporting diversity and providing opportunities to women and minorities. My defined role limited what projects I could work on. I missed the chance to get my fingers into new projects and explore unfamiliar technology. Network security and threat detection was increasingly more important. Efforts to become involved in projects supporting cybersecurity were minimally successful. When an opportunity arose at a company specializing in security software I knew I was going to apply. I had limited knowledge in security applications but I had extensive experience in software development projects, supporting software applications and business analysis. I reworked my résumé and applied for the open business analyst position.

It was almost six years ago when I walked through the door at XYPRO Technology Corporation. I had a lot to learn. The software applications ran on a platform that was foreign to me. The types of applications they developed were different from any I had worked on before. But, everyone was generous with their time and willing to help me as I learned multiple applications and a new operating system. This is truly a supporting and cooperative environment. This is the first organization I have worked for with a woman CEO. I am certain that has impacted the corporate values. There are multiple women in leadership roles and the diverse workforce creates a welcoming and collaborative environment. It is also the first organization I have worked for that really focused on employee satisfaction, wellness and developing new talent. Our internship program has developed a multitude of skilled programmers, security analysts, quality assurance specialists and technical support staff. It is one of the best I have participated in through the years.

As you search for the right fit in your journey through technology you may not start out knowing what you love; it may always be part of your search. But if you are willing to try new and challenging opportunities, always be learning, and trust your instincts then you will eventually land in a place that is right for you. I want to remind you to develop your brand. Demonstrated ability to learn and a good reputation allow you to walk through doors that might otherwise be hard to open.

I will add that I am grateful to all those who invested in me when they did not have to or who took extra time to guide me on my journey. Nobody succeeds at everything on their own. My hope is that I too have helped others find their way through the technology landscape.

I will close with a quote from John Lennon, *"Everything will be okay in the end. If it's not okay, it's not the end."* As I have strived to become a role model and a leader, I cannot think of a better place to have landed than at a company with a strong woman as its leader.

Dr. Gorst's Technology Tips

1. Be honest with yourself that you did not succeed at everything on your own. People helped you. Be grateful for the individuals who reached out to help you when they did not have to. Make sure you do the same.

2. Your character is your brand. When people know they can trust you to keep private what they share with you in confidence, to share credit with team members, and to deliver as promised then you will be the person they think of when they need to deliver and are building their team.

3. You will frequently hear advice to 'do what you love'. It is okay if you do not know what that is. It is equally important to understand what you do not love. Choose classes, projects, or work that interests you. If the work does not keep you engaged you know this work is not a good fit for you moving forward.

4. Be excited, not afraid of tough projects. Clearly understand what details cause the project to be labeled 'difficult'. Ask questions and confirm with the key stakeholder or the project owner that you have correctly defined criteria for what constitutes success for the project. Use this opportunity to expand your skills, your network, or both.

5. Do not wait for people to offer up career advice or mentoring. Reach out to people in your circle who inspire you and make you think "I want to be like her/him". Let someone know how she inspires you

and ask if they have 15 minutes for a quick chat about career options. They will probably say yes. Show up ready to ask one or two questions and honor your time limit. Have your pitch ready to ask if she/he would be interested in periodic discussions or more formally in mentoring you.

6. If someone in a position of authority (manager, professor, guidance counselor) suggests that you should not apply for a position (or school) ask why not. Respect their opinion, honestly evaluate what they tell you but do not let it stop you.

7. Sometimes, no matter how hard you try to force a situation to happen (job, assignment, etc.) it never works out. Perhaps you are trying to fulfill someone else's expectations and not what is best for you. You may be instinctively holding yourself back because you sense you are on the wrong path.

8. Some people are lucky. Many lucky people have prepared by doing extra work, studying longer, or spending their free time learning a new skill. If everyone is successful because they are lucky, and you are not, take an honest look at how their skill set and preparation compares with yours. Yes, sometimes you get lucky. Most of the time luck comes to you because you were preparing for the opportunity.

9. I have seen repeatedly through the years that the people you need often show up in your life when you need them most to be there. The door-to-door salesman who just happened to know calculus was the most peculiar example. I acquired a calculus tutor, and he

remembered how much he loved learning and re-enrolled in school. Keep your eyes and ears open for serendipitous opportunities. They often lead to a new direction or a realignment to your best path forward.

10. It is okay to sometimes be scared, to feel clueless or to be unsure of what to do next. Take a deep breath, acknowledge the situation, decide what action you can do next to move forward and just do it.

About Dr. Carol Gorst

Dr. Gorst completed her Ph.D. and post-doctoral research working at the interfaces between chemistry and biology that are important to biomedical problems as well as the global carbon cycle and the metabolism of methane and CO_2, important greenhouse gasses.

Dr. Gorst focused her technical expertise in the creation of software applications advancing drug discovery research while expanding her understanding of methods in software development and regulatory requirements in the pharmaceutical industry.

Carol brought her 15+ years of experience to XYPRO Technology Corporation with a focus on developing secure, regulatory compliant software applications. She has been essential to establishing processes to enhance the pace of software development, implement customer-centric business requirements, and improve the usability and overall satisfaction with XYPRO's software.

www.linkedin.com/in/carolgorst

Chanel Johnson

"Check your bias at the door and don't use it anymore..."

- Chanel Johnson

Chapter 3

"Be Stemastic"

By Chanel Johnson

I believe my Tech journey started When I was in the 7th grade. It's probably important to mention the type of student I was. I was an average student. I wasn't gifted, I didn't really have a favorite subject. If I liked the teacher then I liked the subject! I was the student who had one to many "Talkative" comments on my progress report. my ELA teacher (I love her until this very day) took our class to the computer lab to show us how to make our email addresses. I was so fascinated with all of the things I could do! My email is no longer active but it was princessdawn@yahoo.com. Hindsight is 20/20, this is not a professional email. That same year my parents bought us our first desktop computer on dial-up! I learned how to use the computer and reach people on the computer. I joined chatrooms, I was the victim of downloading viruses. I joined facebook in 2005, when you could only join with a college email address. I loved technology and all the things I could do and people I could reach. As time progressed through high school, college, and my first year teaching, my technology experience began to strengthen. Imagine my disappointment, my first year teaching in 2009-2010. I started off as an 8th Grade Math Teacher, (Math was my real passion at first) but within a few weeks I was switched to 7th Grade Life Science Teacher at the time. I was unprepared with the content and All I had in my classroom was an overhead projector; a flatscreen TV screwed to a rolling cart. The teacher down the hall had a smartboard. It was the only smartboard in the school. I wanted one so badly! I hated the overhead projector; it was hot, the marker always ended on my hands and clothes. **Most importantly, it separated me from my students**. The overhead did not allow me to connect with my students the way I had hoped. I have a problem. I can't have the fancy smartboard. I hate this overhead projector. What can I do? When a problem is present the opportunity for innovation is born. The flatscreen tv! All I needed was a VGA cord, my personal laptop (my laptop Microsoft office was a version higher than the district laptop), Powerpoint, and a clicker. From this point, my fancy homemade, standards-aligned slide decks were engaging my students. My students loved them! They engaged with me and more importantly engaged in the content. Using slide decks was not my only strategy, it was the start to me seeing positive effects of aligning technology to my content. I remember being in my mid 20's on a Saturday night creating slide decks for my students because it brought me pure joy and I wanted to use technology to relevantly teach and reach my students. At the time, this was considered "new" for us. If you ask another teacher somewhere else they may tell you they had all of these fancy tools and devices in year

2009-2010. That brings up the conversation of equity and accessibility within communities AND even subject areas. That is a different conversation!

Relevantly teaching and reaching MY students was my driving force that landed me many opportunities to come. After my first year teaching I proved myself in ways that leaders were expecting, classroom management and high scores. As new technology arrived in my school, I was one of the first to have access to them. The year all teachers received Promethean Boards, many of us did not know how to use them, I would stay after work for hours watching youtube videos and calling the Promethean help desk to get assistance on best practices. I saw first hand how using technology in the classroom could engage students so I had to master this! During my third year of teaching I was teaching 8th Grade Physical Science. This is pretty much where I stay during my years in the classroom. This particular science gave me a balance of math and science in ways I didn't know that I loved!, 2 major instructional strategies were on the rise. 1. LMS (Learning Management System) 2. Interactive notebooks. I chose to focus on an LMS, specifically Edmodo. Now in my room, I was full of technology (desktops, Promethean boards, laptop carts, Ipads, activ slate, activities activ expressions. This was a major change from my first year teaching. I was thankful to have administrators that saw the effectiveness in my teaching with technology they did what they could to make things happen for me. At the time many of my colleagues did not know how to use technology or were not interested in using it. I remember a colleague coming in to see my "tech funhouse as she called it" and asked me if I could help her. I didn't know that helping a colleague was opening another door for me.

Have you ever heard the phrase, I go in my classroom, close the door and do me? That is pretty much what I did until a colleague asked for help. I was able to train her on what she asked and immediately she tried it out and felt empowered. Her feeling of empowerment empowered me! More of my colleagues began to ask for help, in-house that is what I did. The same satisfying feeling I felt when I saw the light-bulb turn on for my students was replicated when helping my colleagues.

My principal at the time introduced me to GAETC, Georgia Educational Technology Conference. I was so excited to go because I was going to learn so much. When I went to this conference I left there feeling inspired. I learned a lot and saw presenters that were normal educators like me. I wanted to present, but didn't think what I had was worth presenting. I let this feeling go for a while and continued in my classroom and supporting teachers in my school.

The year I presented at my first conference, I presented at GAETC. My topic was "Flipping your classroom using Edmodo". My room was a nice size and wasn't full but it was a nice enough number. Regardless, anyone that stepped in that room was going to hear some good music and get the best of me. I hated the feeling of anxiety I felt before my session but loved that rush after I was done. My favorite part was connecting with the educators afterward. I remember one educator asked me if she could connect with me on Twitter. Twitter?! I didn't have one. At the time I was not into Twitter and laughed at anyone that had one. The impacts I was making in science were also gaining attention which afforded me opportunities. The district had an educators team called " The Vanguard Team". The team included educators from all over the district.- They worked together to learn best practices, teach our colleagues in house how to effectively use technology to transform instruction, had fancy tools and swag. I knew I had to be part of the program because it fit everything about me. I applied and did not make it! I never accept a rejection without an explanation. I asked why, not to be defensive, but to see what was missing and how to be better. My explanation was that I wasn't showing how technology was used in transformative ways. In my defense, I felt reaching my teachers where they were and growing them was transformative. Needless to say, I felt hurt that I wasn't good enough for that team, but happy I was supporting my teachers.

My last year teaching 2015, I went out with a bang! I was the teacher of the year that year. By summer I had 2 job offers. I could be a Technology Coach or I could be a STEM coordinator/ Admin Assistant for a middle school. I chose the position that aligned with my passion. My passion lived in STEM education. I selected STEM coordinator. This was my first year out of the classroom and into an evaluative role. This was my most challenging year, I left my home school. I knew the people and the culture there. I was no longer "us" I was "them". Teachers, you know what I am talking about. I had to learn so much. A new school, the culture, the people, the expectations on paper did not match the actual expectations, and how to be a leader. Going from close my door and do me, to being in everyone's classroom was truly a shift. Growth is not a comfortable feeling and this was that year. I was able to create STEM programs and work with district leaders to provide resources to our school to help support our programs. One of our programs focused on Adobe Suite. Adobe was only funded for high school programs. We were able to get an exception from the district. We started the Junior Level TSA program and started a coding program. This experience taught me how to partner with others and to specifically ask for what I needed. While getting the hang of this job, I presented at more conferences. The opportunity to apply for the Vanguard Team came up again. I applied, took the feedback from the first time, I was accepted. Not only was I accepted, but the Vanguard leader for my school. Vanguard opened so many doors for me and I connected with so many educators who were liked minded in my district.

The opportunity to show my skills and reach people finally occurred at a conference hosted by our district. I twisted my ankle coming in the door, feeling defeated. Limping and walking it off, I presented about Edmodo. This session was a full house! We had to get more chairs. I remember looking into the audience seeing teachers from my old school, teachers from my new school and teachers I had not met yet. After the session, a teacher came to me and asked me "do you have a twitter?" The answer was... no. It was my last "no" that night I created a twitter account. One of the first companies I followed was Edmodo, I wanted to share with them what I was doing and how I was using the product. This afforded me the opportunity to become one of the Ambassadors for Edmodo. I also began connecting with educators on twitter opening my eyes to a world outside of my school district and even my state. I began participating in Twitter chats hosted by our tech team. And chats around the world. Twitter chats were about a topic. The host will ask a question with the format Q1 (Question 1) and the hashtag to the chat. The participants respond A1 (answer one) and the hashtag to the chat. The chats gave people insight on who you were and how you think. From there educators will connect with you. I use to commit to about 3 a week, either as a host, participant. I truly believed the best professional development I ever had was done in my Pj's!

My last and only year as a Stem coordinator and now member of the district Vanguard Leadership Team presented an opportunity to be promoted as a district level employee. I was now a STEM specialist with a focus in K-12 Math and Science. I supported 18 schools. I remained a member of our Vanguard team which grew from 200 educators to close to 500 educators. When I joined the Vanguard Team I started as a leader in the school to learning community leader and eventually one of the executive leaders of the entire organization. Being part of the Vanguard team was volunteer, while doing my actual job as a specialist. The beauty of this was, I was able to connect technology to instruction. I could reach more educators, learn from more educators. Being a leader gave me the chance to attend the largest technology conference of all, ISTE. This is where I connected with so many new edtech educators. Lack of people of color in the edtech space was an issue. The few of us that attended ISTE all connected and remain connected. I was the new kid on the block in a world I didn't know existed. Connecting with these educators invited me to be part of a movement of advocating for people of color in edtech spaces. Every year after my first year at ISTE, I became a presenter at the conference, a member of the Digital Equity Group where we focus on tech access and opportunities for all. Everything seemed to be going well in my tech leadership journey. I was presenting about what I loved, applying it to practice, surrounded with mentors and continuing my learning. The one hurdle and challenge I faced was getting educator buy-in (teacher, principal, coach) to wanting to integrate tech tools in the classroom. Some did, but I wanted all to do it because it was what was best for our students. I didn't want to see teachers using 1:1 devices just to do worksheets. I wanted to see students create, I wanted to see students using simulations to apply scientific concepts, I wanted them to be producers of content not consumers. At the time it wasn't viewed as a necessity until it was.

The Pandemic Hit like a thief in the night. It changed the way we did school and is still changing. The Pandemic turned every instructional technology coach into a superhero overnight. While I lived in both the Content and Technology world. It really narrowed my focus on ways I could support and lead others. I became ISTE certified during the pandemic. This was one of the best certifications I joined!

My Masters is instructional technology with an endorsement in online teaching. On paper this sounds like I was prepared for the pandemic. I received this degree in 2015. So much has changed since then. That is why it is so important to remember the degree isn't the end game, you still have to continue learning and improving your practices. Especially in the technology world whether you are corporate or education. Going through the ISTE certification was the test I needed. I actually failed the first time. They say failure means First Attempt In Learning, that's true only when feedback was included. I am thankful that I knew "how I failed" where I lacked and how to get better. This certification was a true test to my practices and my ability to communicate my practices.

Today I am fortunate to connect with thousands of educators, I am constantly learning and improving my practices. My favorite instructional model that I base my work off of is the TPAK model. It is the overlapping of content knowledge, pedagogical knowledge, and technical knowledge. It is this model that makes it difficult to truly define me or attempt to put me in a box. My focus isn't just science, my focus isn't just technology, my focus isn't just good instructional strategies. We can talk about these pieces individually, but I truly believe you won't get my best if I do not talk about them all and how they all intersect. I am featured on multiple podcasts about different parts of my work. I even created my own mini workshop about engaging scientists usingGoogle Jamboard for teachers all around the world. I remember this teacher told me "it is 3am where I am and could not miss this"! I felt so honored.

Chanel Johnson's Technology Tips

1. **Learning Goals:** This is probably my most important tip of all! You can have all the tech tools in the world. None of that matters without a learning goal. The learning goal always comes first. Learning goals can come in many forms. I like to start with the instructional standard or building blocks that lead to the standard. Instruction comes first! This will help you when it comes to the "why." Why am I using Google Jamboard? Why am I using Nearpod? Why am I using Flipgrid? What is the learning goal I expect my students to master? The end goal is student achievement. The learning goal should be the "why" you are using the tool in the first place.

2. **Pedagogy:** This is the "how" what instructional strategy are you using? Is the strategy whole-group instruction, stations, collaboration? Often the strategy is in the standard. For example, in Science, the standard says, "Develop Models to identify parts of a plant. Developing Models is a science practice but can also be used as a strategy. Students are developing models. What tools can I use that allow students to develop their model? Students are creating, so think of a tool that will enable students to create.

3. **Engagement:** I recently conducted training about Google Jamboard in Science. I started the training by asking 300 teachers what "Engagement" looks like? I used Mentimeter's word cloud feature.

If you know how a word cloud works, the word mentioned the most would appear as a bigger word. There was so much rich insight I gained from this activity. 1. So many different answers define "engagement," yet we put such heavy emphasis on this word in education. We don't have a consensus on what it means. What do you think the big trending word used to define engagement? Participation. For this purpose, I am using the Center for Applied Special Technology's explanation of engagement. It says there isn't one means of engagement that is optimal for all learners. We have to provide multiple options. The tip here is to have various tech tools up your sleeve that hits different modalities so students can demonstrate their learning.

4. **Be Receptive but Protective:** My previous tip ends with having multiple tech tools that hit different modalities, which brings me to my current tip. Stick to 3-4 useful edtech tools that can be used for different things. Master those instead of trying to keep up with 100 various tools. New tech tools are born every hour. Many do the same thing that your current tool is doing. Be open to learning something new, but have self-awareness regarding having the capacity to take in a new tool. I am excited about learning the latest tech tools, but even then, my go to tools are Flipgrid, Phet simulation, Nearpod, Google Jamboard, Microsoft Teams, Canva, and Microsoft One Note. I use many other tools and like them. When I feel the need to switch it up, I will.

5. **Stay Connected:** Social media is more than just posting cute

pictures, updating your family and friends about what's going on in your life. When I use social media, I am using it to connect with educators all around the world! Some of my best professional development occurred during a Twitter chat, being in a Facebook group about a specific topic, or listening in on Clubhouse. Staying connected has afforded me so many opportunities allowing me to connect with so many people. Dr. Sarah Thomas, a dynamic educator, person, mentor, and current author in this same book, talks about building a Personal Learning Family! This occurs by staying connected. If you are just getting started. Follow me on Twitter @Dc_stemtastic and IG Chanel_the_steminist

6. **Build your Soft skills:**Technology has so much to offer, so many tools that can do so much. I recently transitioned from a PC device to a Macbook. The learning curve was tremendous to me. I didn't want to use the device anymore. Everything was different, Windows had ctrl, and Mac had command. Windows closes a browser on the left; Mac closed a device on the right. I wasn't ready! I printed my Mac keyboard shortcuts, began to practice. I am using this as a real-life example to encourage you to take the time to learn soft skills such as shortcuts on your device, and how to split your screen. It can make your work faster.

7. **Check your bias at the door and don't use it anymore.**I had the privilege of doing my job from home, supporting my kids during virtual learning. My kids can create the coolest Tik- Tok video. They have phones, laptops, and tablets. I assumed since they have all of

those devices, then virtual learning would be a breeze! My kids struggled with navigating google classroom, splitting their screen, finding the different shortcuts I mentioned in Tip 6. I assumed that since my kids could do all of these fantastic things, they could do simple soft skills. I hadn't taught them those skills. These skills were not skills they thought to have, so of course, they didn't teach themselves. My tip here is don't assume if a child has the device that they can do all things you ask of them. Don't think a child who may not have a home device lacks tech skills to be successful. I see many struggles in virtual classrooms because students need the soft tech skills to navigate to the assignments. This could be from the bias that the younger ones automatically know what to do.

8. **Digital Agency:** I remember having a conversation with someone in my PLF on Twitter about the word "digital citizenship" the bias and potential negative undertones the word "citizenship" could have. At the time I didn't read into it that way. I looked at citizenship as a status and agency as ongoing work. Being a digital agent is using technology responsibly. Common sense education has resources to guide your journey towards using technology safely. Developing young Digital Agents Is the responsibility of both the parent and the teacher. I will give you an example. I created an Instagram account for my kids. They do not know the password. I monitor it. We model what examples of a positive post. We look to evaluate what negative posts are. Kids are out here with social media accounts. Some have the "secret" account. It was vital for me to teach my kids how to

navigate through social media and how a bad post can come back to get you. Digital Agency isn't about me telling you what not to do, but providing you with the skills to make good decisions during your digital journey.

9. **Permission to Fail:** Having all the skills, a plan A-Z for troubleshooting still does not guarantee a Tech-tastrophe won't happen. Think of your Tech-tastrophe? This is when your tech messes up during the most inconvenient time. My Tech-tastrophe was me presenting during a PD. Everything is going well until my computer shuts down due to me forgetting to charge the laptop. That day was a fail! Days like this or maybe a day where your tech just didn't work (that happens), it's okay. Fail Forward!

10. **Growth Mindset:** Have you ever said to yourself or heard someone say I am not tech-savvy. I am not good with technology? The moment you fixed your mouth to say this out loud or think it in you, you have put yourself in a box called the "fixed mindset" this box has a top on it, and it closed shut. Take the lid off of that box and open it up! Having a growth mindset in tech is where success can live. Challenge yourself to learn a new tool, take a risk, work through that tech-tastrophe.

About Chanel Johnson

Chanel Johnson is a native of Atlanta, Georgia, where she currently resides and serves as a Science Specialist for the fifth largest school district in Georgia, Clayton County Public Schools. She has served countless students and educators since 2009, focusing on the areas of math, science, instructional technology, and coaching. Whether teaching students or teaching adults, her passion is to teach and ignite all voices so that they are heard. She believes in developing digital age learners and digital age educators by stimulating a culture of empowered learners through applying 5 C's (Creativity, Collaboration, Citizenship, Communication, and Critical Thinking). Her passion is to show stakeholders how with current researched-based instructional strategies, empowering all learners' voices, especially those often silenced, fostering a dedicated community, all learners can meet their highest potential.

Chanel usually develops and executes professional development around the implementation of content and instructional technology fluently. She believes instruction consists of Standards-Aligned Content, Appropriate Pedagogy, and Technology Integration. The most crucial factor is to remember we are teaching the child before the content.

Chanel believes in the power of community and collaboration through being a part of countless professional learning networks often sparked from social media platforms such as Twitter. You can follow Chanel on

Twitter as Dc_stemtastic. She has conducted several professional developments in training coaches, educators, and instructional leaders towards using the most appropriate technology to empower all voices in science education. She has presented at countless national and local conferences, served as an ASCD's Emerging 2019 Leader, MIEE, Flipgrid Ambassador.

Chanel is a proud wife to her husband, Martin. Mother to three kids, Shaylyn, a set of twins, Christian, and Chloe.

Dr. Jocelyn McDonald
"Dr. Mac"

"Leverage technology as a vehicle to make a positive impact."

-Dr. Mac

Chapter 4

The Road to #TechItUp

By Dr. Jocelyn McDonald

My journey to technology began with an interest and love for science as a young girl. I was so captivated with learning how things worked and problem-solving, that it was not until later in my adulthood that I realized that all roads pointed me to technology as a passion. I have always excelled in math and science throughout my primary and secondary years in school and was often recognized by my teachers for my accomplishments and hard work. It's amazing how teachers play such a subtle but major role in their students' lives. Although there were many teachers who supported me, there were three, to this day, that may not know how much they have shaped my interest into the work I do today. It started with my elementary science teacher Ms. Barwell who shared her passion for the love of science beyond our classroom walls. Her positive energy and support continued to drive my interest in science all the way to high school where I had the pleasure of

Mr. Henderson being the first person to expose me to computer science and learning to code. Then my chemistry teacher, Ms. Mason, who publicly recognized me for my accomplishments in her chemistry class. So, it was a natural decision when I decided to go to college to double major in chemistry and computer science. It was very challenging to master both chemistry and computer science courses at the same time and I often found myself working late in the science lab and in the computer lab. After balancing many classes at the same time for both majors, I decided to focus on chemistry and planned to attend graduate school after I earned a Bachelors of Science degree in chemistry and become a Pharmacologist.

The Unplanned Career in Education

During my last year in college, I quickly found out there was another plan for my career. I learned that I would be expecting my first child and naturally a whirlwind of uncertainty arose. My son was now my new priority and also my motivation to finish my degree as I had become a single parent. Being a single parent shifted my focus and had a major influence on my career path. During my pregnancy, I continued to take evening courses and worked full-time as a substitute teacher. My experience substituting began to shift my direction from wanting to become a Pharmacologist to now becoming a chemistry teacher. However, teaching was not an original passion of mine but developed into a slow metamorphosis of my calling.

The Hustle Road of Entrepreneurship Leads Back to Tech

My first year in the teaching workspace was not as simple as I thought it would be when I was a substitute teacher. It was challenging for me to teach a complex topic like chemistry at the high school level to students that I was barely older than. In addition, I was still going through life trying to figure out my purpose and make more money to provide for my son. At the time, I did not see education as a longevity career, so I began to look for other opportunities to make extra money. I was introduced to network marketing and found an interest in becoming an entrepreneur. Of course, starting any business requires some form of capital to get started and continued funds to run and manage the business. My finances were limited on a teacher salary, so I began to leverage my love for learning, problem-solving, and gain skills to get the most out of a cost-effective budget to support my business. This road led me to leverage technology to support marketing and design. I began to self develop and become more knowledgeable by attending seminars and workshops on creating sales copy and internet marketing. I found myself applying my coding skills, using HTML, and learning to leverage technology platforms to promote my business. I also found myself re-inventing my image as an entrepreneur which then trickled into real estate investing and applying these same skills to market myself. Never in a million years did I think I would apply this enhanced skill set to education as it was my goal to exit out of teaching. And I was gung ho on a mission to ensure my business would be successful until another life turn derailed my drive.

The Hell of 2012 and Reflective Thinking

The year 2012 was a very memorable year of loss and reflection. Although my business was finally taking off with a flow, the loss of my grandmother was very devastating which also followed by multiple deaths in the same year. My grandmother and I had a close relationship and I began to reflect on my work and purpose in life. The loss of her allowed me to reevaluate my life. I decided that I wanted to fulfill a promise I made to her as a little girl and become a doctor. She was Shirley Mac and I would tell her growing up that I would become Dr. Mac. Throughout the hustle for entrepreneurship, I was still teaching chemistry, mainly in low socio-economic schools, in which I was looking for innovative ways to connect learning with my students. What I didn't realize was there was a force reeling me into my work as an educator from which I built relationships with my students in a way that made me feel like I was making an impact on a young person's life the way my teachers had on me. That feeling was very rewarding and felt different from just making money to leave a profession I did not realize I would become passionate about. At this point in my life, I had been teaching for quite some time and I was still developing as an educator. Although I realized teaching was rewarding, I still felt burnt out by the end of the day and knew there had to be a better way. This then led me to slowly start implementing technology into my instruction to help maximize my time and be more efficient throughout the day. After all, it was a year of reflection for me. I decided to develop myself and go back to school to get a Master's degree so that I could eventually get a doctoral degree. Naturally, since I was already working in education and found myself using technology in a way I had not before, I earned a Master's in Education degree in Curriculum and Instruction specializing in Instructional Technology.

Back to School, On a Mission, and Yet Another Restart

As I began to try new things and further develop myself, I quickly saw the shift and impact I was making with students and thinking outside the use of technology. Some of my challenges during my transition was school culture and awareness of the technology. I found often that school administrators did not understand the concept of 21st-century skills students need to be successful to be productive citizens in our society. I found there that more focus for student success was put toward standardized and normalized testing. There was no culture centered around leveraging technology as a driving force to support instruction. The technology was only looked at as a tool and not a vehicle for change. I often felt that my thinking may have been too forward for practices that were supported at the time and probably still too forward for many schools until the Pandemic of 2020. However, it was very important for me to continue developing myself to become a scholarly expert and complete my Ph.D. in Education Technology. In addition, it was also important for me to be an educated woman of color to model for other members of the black community that poverty does not define you or your abilities to be successful. When I go to schools, I want young girls of color to be able to see themselves in me. It was important for me to exemplify black excellence and that someone who was seen as the pretty girl could also be the smart and driven girl that set and attain goals.

However, life has a funny way of thinking that it can hold you back and throw new curveballs. At the start of my doctoral program, I found out that I would be expecting another bundle of joy. At this point in my life, I experienced so many things that could have derailed me and I had learned to push forward and not to let circumstances define or prevent me from achieving goals that I set for myself. My son was now 12 and I found myself starting another cycle again with my daughter and once again a single parent. Of course, I could have easily fallen into a slump, but I realized that life is full of challenges and I was determined to fight through those challenges because I also know there is always a brighter side. Now, I am even more motivated to push through and take on one step at a time. Sure bumps in the road continued, but I knew they were only temporary roadblocks and I understood how I chose to handle them would determine my success at moving past them.

Giving Myself Permission to be Great!

When I finished my doctoral program and earned my Ph.D. in Educational Technology, I felt accomplished. Not only had I made a groundbreaking success through many barriers in my life but I was also overjoyed and filled with confidence. The entrepreneur in me was still thriving but now with more time and purpose. I started transforming my entrepreneurial efforts into my passion as an educator and applying that skill set to the work I was already doing in education. My work as an educational technologist for a large, urban school district was only a layer of my goals and vision for myself. My career in public education has allowed me to help and develop so many people that I was now wanting to extend beyond the political nature of public education and make more of a positive impact in our society through the use of technology. I want to be a voice for change in education to work toward barriers that have a positive impact on equitable access to effective learning environments to develop our students for the digital world we live in as well as make an impact on others' livelihood and well-being. Also, I saw a need to support the digital divide that exists in specific communities and a way that I can contribute to support African-Americans and lower socio-economic communities through my work in technology. As an African-American woman who has propelled through generational poverty, I am a strong supporter and believer of developing digital literacy and technological skills in low socio-economic communities to support the advancement and culture of others as well as break the cycle of generational poverty.

To accomplish this, I had to "give myself permission" to be great. Les Brown, one of my favorite motivational speakers I often listen to, always famously shares with his audience "You have greatness within you." The power of this statement has helped shape my thinking, my ambition, and my drive. It is important for me to do more than tell myself I am great. I had to embrace it and believe it to push forward to my next calling to be that voice that makes a positive impact. I am a firm believer of going beyond the status quo and that any work I am connected to has to be done with purpose and stand out to make the intended impact. The essence of my brand, #TechItUp, then became anchored into these three words: Lead. Inspire. Develop.

Leading to Inspire

To take initiative in leading to inspire, I decided to start a technology blog to support others' use of technology and address gaps often observed and experienced in education. I finally took a step forward with my ideas and releasing the greatness within myself to be shared with the world. This then led to me stretching myself even more and starting the #TechItUp Talk Podcast where I discuss challenges in education and provide tips and best practices to integrate technology to support educators. I then began developing workshops and professional development opportunities for educators. I also created the #TechItUp Digital Skills & Technology Student Camp series to inspire and develop student proficiency in digital skills and applications to support the online learning experience and prepare for real world applications of technology into our society. As one opportunity led to another, I found myself thriving in opportunities, developing educators, and consulting for professional learning, to supporting small businesses and entrepreneurs with integrating technology into their business model to increase their digital presence, productivity, and efficiency.

Although I finally began to feel the rewards from the positive impact I was making by leveraging technology as a vehicle, I also realized there were not many educational technology leaders or even in the general technology industry that looked like me, an African-American woman driven to lead and make an impact in technology. Now with more angst, desire, and passion I look forward to continuing to break down barriers to create more opportunities for myself and others. It's an amazing feeling to let your passion to support others drive your ambition. Then your ambition can then create opportunities beyond your imagination.

Dr. Mac's Technology Tips

1. Get more out of multifunctional platforms to enhance efficiency, productivity, and organization. These may be existing applications that are readily accessible such as Microsoft Office or Google Suite and other web-based technologies.

2. Educational institutions and organizations should optimize the use of instructional software and applications to create positive learning environments that will maximize instructional processes and increase engagement with learner-centered strategies.

3. Organizations, small businesses, entrepreneurs and/or edupreneurs do not need to have a Fortune 500 company budget to integrate technology as an asset to operate and manage their business or processes. These entities can easily get the support they need to build capacity and align their organization or business goals to technology solutions that support efficiency, service, and productivity.

4. Leveraging technology consultants are cost-effective ways to help organizations and small businesses align goals with technology solutions that will support the implementation of their productivity and strategies to service their clients.

5. Use technology to automate your systems to strengthen workflows, communication, data, and social media presence.

6. Create a digital presence that sets you apart from others. It's essential to think about how your digital presence stands out from others and ask yourself is it consistent and/or prolific?

7. Create a technology plan that supports the mission, vision, and goals set by the organization to develop digital practices and processes that enhance the functionality of your systems.

8. Find opportunities for professional learning that will support the integration of technology into various educational and business models.

9. Leverage virtual spaces to capitalize on digital learning experiences, communications, customer service, marketing, and even product distributions.

10. Learning institutions and businesses should create a culture where technology is not an additional resource but an integrated partner throughout systems, instructional delivery, and daily functions.

About Dr. Jocelyn McDonald

Dr. Jocelyn McDonald is an Educational Technologist, Podcaster, Blogger, and Technology Integration Consultant for professional learning and small businesses. She is a proud advocate for social change in education and is dedicated to making a positive impact. She holds a Bachelors in Chemistry, Masters in Curriculum and Instruction in Instructional Technology, and a Ph.D. in Educational Technology. With 15+ years in education, she is passionate about supporting the transformation of education through technology for equitable access to 21st century education.

Website: DrJocelynMcDonald.com

Email: info@DrJocelynMcDonlad.com

Twitter: @DrJEMcDonald

Instagram: @DrJEMcDonald

LinkedIn: @DrJEMcDonald

Facebook: @TechItUpTalk

Karen Walsh

"Give back what you received."

- Karen Walsh

Chapter 5

An Unexpected Journey

By Karen Walsh

Looking back on my journey to the cybersecurity space, I can see the foreshadowing sprinkled throughout my life. I was an early adopter before people knew that early adopters existed. Growing up in the 1980s and 1990s, computer ownership was the exception, not the rule. Computers had just started moving into homes. As I vaguely remember, my family's first computer was a hand-me-down from my uncle, an engineer for a large aircraft company. The more time I spend talking to cybersecurity professionals, the more I realize that my history aligns with theirs. We started out with the old school DOS personal computers. We were the kids playing *Where in the World is Carmen San Diego* in the 1980s and adopting America Online (AOL) chat rooms in the 1990s. We adopted LiveJournal and Friendster long before WordPress and Facebook. We personalized MySpace pages and chatted over ICQ. When I finally joined the cybersecurity world, I realized that I had found my people.

The Long and Winding Road to Cybersecurity

Unlike hackers who started coding young, I never considered technology or cybersecurity a potential career path. I did the traditional late-1990s liberal arts career path. After graduating from college – in three years and as a Phi Beta Kappa inductee – I worked in the insurance industry. The department that hired me had a tuition reimbursement program that paid for part of my law school degree.

After passing the Bar exam, I decided to start my own company, Allegro Solutions. The company's first iteration focused on providing part-time compliance services for community banks. From there, it grew to engage in part-time internal audit functions. To supplement my business, I began teaching first year college writing, a career choice that changed my life.

The eleven years I spent in academia paved the way to my role as a content writer in cybersecurity. Even though I always logically understood focusing writing to a defined audience, teaching reinforced that and added in the need to meet my students where they are. Students uprooted some of my long-held beliefs around the knowledge they brought with them from high school. Over the eleven years, I learned that sometimes the assumptions we hold about our audience's knowledge are wrong. This realization is one of the reasons my cybersecurity writing business is successful. In the information security industry, we need to speak to various cybersecurity awareness levels.

In 2016, my audit career came to a grinding halt as my last client decided to stop engaging in audits. I advertised myself on Upwork as a compliance expert, and a cybersecurity compliance platform hired me to do their content. Since 2016, I grew my business, moved to an in-house product marketing position at an Identity Governance and Administration platform, then went back to my business in September 2020.

Being the Change I Want to See

Although the number of women working in cybersecurity has increased since 2016, we still remain a small overall percentage of the workforce. In many ways, this is both an asset and a liability. As a cybersecurity marketing content writer, I find myself working across the business and technology sides of the house. Marketing assets seek to communicate value across both of these areas. For a writer to excel, she needs to have a fundamental understanding of both the senior leadership needs and the technical requirements. However, as a successful business owner in this space, I also have the opportunity to be the change I want to see in the industry.

Cybersecurity still suffers from gatekeeping, asking women to prove their value while assuming men possess inherent value. Despite being an attorney and published author outside of the security industry, I often find myself questioning whether I belong here. The reality, however, is that the information security industry often needs non-technical minds as much as it needs technical minds. Hackers, both ethical and malicious, excel at programming. They excel at understanding technical risks.

However, many are quick to say that the "end-user is stupid" without realizing that their gatekeeping is the reason the end-users feel inadequate.

Infosec Twitter, where many elite hacker minds congregate, is rife with "why don't people use better passwords?" or "it's their own fault for clicking on links in phishing emails." Problematically, this mentality only adds to the divide between the technical and line-of-business users. As a content marketing writer, I work hard to use my position as a way to influence by educating. After all, buyers need to understand *why* a technology helps them before they can make an informed decision around an expensive purchase. Bridging this knowledge gap is how I built my business from one earning $18/hour in 2016 to one making six figures in 2021.

The more I interacted within the cybersecurity community, the more I realized it was the professional home I had long sought. These are the people who were also the technology "early adopters." They were the ones who had also spent time on LiveJournal or AOL chat rooms as teens. They were the ones who had learned to use computers before we had operating systems, back when you need to use command lines to boot up a computer. These are the people who understood my youth in ways that others didn't and couldn't.

My success is due, in large part, to my desire to educate. I have no desire to code. I have no desire to be a hacker. I have taken some introductory lessons, here and there, so that I can more effectively communicate industry issues to business personas without writing marketing copy that increases technology experts' skepticism. However, at my core, I remain an educator trying to be the change I want to see in the world.

While that might sound like a lofty ideal, the reality is that my success as a technology writer comes from my educator background. People buy technology when they understand how it helps them achieve their goals. According to one report from June 2020, experts valued the current cybersecurity market at $156.5 billion in 2019, expecting a compound growth rate of 10% between 2020 and 2027.[1] The proliferation of solutions means that buyers often find themselves struggling to understand the differentiation points. In fact, my skills directly enhance the industry because I am not a programmer. Programmers build the technologies. Educators and writers explain why the technologies matter.

As an expert in cybersecurity and privacy compliance, I bring diversity to a space that often lacks it. As a woman, I have both the ability and obligation to make this space safe for others to join. As a writer, I communicate messages beyond a marketing "call to action" by focusing my writing on the most immediate problems companies face when trying to protect themselves and their customers. As an educator, I always view my content as a way of boosting cybersecurity awareness, even if done

---

[1] Grand View Research. (2020). "Cyber Security Market Size, Share & Trends Analysis Report By Component, By Security Type, By Solution, By Service, By Deployment, By Organization, By Application, By Region, And Segment Forecasts, 2020 – 2027." Grand View Research. June 2020. https://www.grandviewresearch.com/industry-analysis/cyber-security-market

through marketing content. All of these have made me a leader in this space.

Although I rarely view myself as a leader, I do know that my commitment to my core values and my approach to my work makes me unique. For example, as my reputation in the industry has grown, I find myself responding to media requests, helping reporters explain how a data breach occurred or why a security vulnerability is important. Again, my approach is to add value, grow knowledge, and help make readers more aware of the reasoning underlying an event. To do this, I need to be able to explain complex ideas, ones that often feel abstract, in concrete, approachable ways. I want to be the change I want to see in this industry by focusing on mind sharing and bringing people together.

Security Through Clarity

Security professionals often argue "security through obscurity," that hiding sensitive information within code protects from attacks. However, I argue that we need to focus on security through clarity, that making cybersecurity approachable to end users is equally effective. As a thought leader in the cybersecurity space, I firmly believe that protecting today's data against tomorrow's threats can only happen when people, end-users, understand *how* to protect themselves. Obfuscation in programming secures code, but clarity in content secures the increasingly important human element.

Karen's Technology Tips

1. **Never give up**. You're going to fail. Everyone fails. Cybersecurity is an industry where failure is more of a "when" than an "if" because cybercriminals are always one step ahead of those protecting data. The important lesson is to get back up, dust yourself off, and start all over again. " Karen Walsh Founder, CEO Allegro Solutions

2. **Be your best self.** The people in cybersecurity who matter will always respect someone who brings her best self to the conversation. Being respectful of others, more than anything else, is the way to build your reputation.

3. **Give back what you received.** The cybersecurity community is a community. If you've received help, make sure to give back just as much as you got. If you're technical, share that with those who want to learn. If you're not technical, offer to help people hone their non-technical skills. I've given online writing tutorials and responded to threads about how to bridge the language gap between technical and non-technical leaders.

4. **Find your niche**. Cybersecurity is a broad term encompassing a variety of different areas. You can focus on external vulnerability detection, dark web research, identity and access management. No matter what you're interested in, you'll be able to find something you love

5. **Ask questions**. More than any other industry I've been in, cybersecurity is one that relies on people asking questions and being intellectually curious. If you don't understand something, ask. People will always help you out if you're respectful and polite.

6. **Take risks**. Since the cybersecurity industry is constantly evolving, you have the opportunity to take risks. Jump into a conversation that you think you might not be qualified to join. Join a technical organization and learn from those around you.

7. **Ignore the haters**. In the same way that you can find your crew on the internet, you can also find yourself subjected to vitriol. Then again, if the trolls are picking up on your voice, you're probably doing the right thing and have a group to support you.

8. **Find your people**. If you're a digital person, you can find support on #infosec Twitter or just by joining online forums. From there, you can grow your network.

9. **Read everything you can find.** Because cybersecurity deals with protecting cloud-based data, professionals share a lot of work on the internet. It's definitely a field where, in a lot of ways, most things you need to know you can find on the internet.

10. **Never feel invaluable.** Even if you're not a technology person, you offer something valuable to the community. For example, I bring the ability to marry technical language and business level understanding together.

About Karen Walsh

Karen Walsh is a data-driven compliance expert focused on cybersecurity and privacy who believes that securing today's data protects tomorrow's users. Karen has been published in the ISACA Journal experience in cybersecurity centers around compliance. Her work includes collaboration with security analysts and ghostwriting for c-suite level security leaders across a variety of internal and external vulnerability monitoring solutions. As a lawyer, she is deeply knowledgeable about security and privacy laws and industry standards including GDPR, CCPA, and ISO. She is currently under contract with Taylor& Francis and is writing a book about cybersecurity for small and midsize businesses.

Laura Hart

Be ready to pivot and embrace the process.

-Laura Hart

Chapter 6

Pioneering STEM Education In New York City

By Laura Hart

When I was very young, my dad ran for mayor of Albany, New York, our hometown. He lost, and in retrospect, he was tilting at windmills, the chance of his victory was very small. Watching my father run for office and lose the election led me to think big and to reach for my own larger-than-life goals. I became comfortable with the knowledge that my goals and plans may not happen easily or quickly.

Throughout my teens, I always worked with kids: camp counseling, being a shadow for an autistic student, and volunteering within elementary school classrooms. **I loved thinking about how we learn and I loved teaching**.

Then, during my undergraduate studies, while I came into my own as a painter and sculptor, my interest in STEM (Science, Technology, Education, Mathematics) took root. At my internship with the Capital Children's Museum in Washington D.C. I attended a talk on Children and Learning by a man named Dr. Seymour Papert. Little did I know that that first meeting would change the course of my life.

In his talk, Dr. Papert, a student of Piaget's and the founder of the MIT Artificial Intelligence Lab (what later became the Media Lab) shared a programming language he had developed called LOGO, but at the heart of it all, his talk reignited my interest in how we learn. I returned to Skidmore eager to study LOGO and learn more. Back then no one knew what I was talking about but through a series of independent studies with one of Papert's students, I was able to learn about technology, children, and LOGO. I was an irreverent school goer who loved to learn, but disliked the school system, and as such, Dr. Papert's work spoke to my soul. It helped clarify my thoughts about how we learn, don't learn, and how we can set up optimal learning environments for children.

Upon graduating and beginning my teaching career at Buckley (where I subsequently taught one of our President's sons), I had to design my own curriculum for Kindergarten-9th graders. There was no computer program when I arrived at Buckley, so I went from a blank canvas in art school to a blank screen on a Commodore 64. I had taught myself LOGO because there was no Educational Technology field at the time. And now I began to design meaningful, engaging projects that married creativity and programming.

While teaching at Buckley (and later The Little Red School House), I spent my summers painting in Maine. By some coincidence or a guiding hand, in 1986, I literally ran into Dr. Seymour Papert in a bakery in Blue Hill, Maine. This chance encounter led to a collaboration with Dr. Papert on teaching children and teachers.

Some people come into our lives for a short visit and have little effect. Others appear and reappear unexpectedly changing our lives forever. That's how it was for Seymour and me. For over 30 years, Seymour was my mentor, business partner, and friend. When I first saw him in the bakery, I was not sure if I should speak up. After all, I was not trained as a computer scientist. I was completely self-taught. And here I was meeting the person who went to LEGO and developed both the programming language LOGO and LEGO Robotics! I soon discovered we were kindred spirits and that he didn't often have an opportunity to work deeply with someone outside the Ivory tower of MIT, someone on the ground using his work.

How is it that I began a business teaching STEM before STEM existed? Seymour Papert inspired me. How did I begin teaching coding before the word coding was used by laypeople? Seymour challenged me. He didn't directly challenge me to do exactly what I did; he challenged me to follow my instincts and **my yearning for a safe space for children to discover their own creativity, courage, and abilities**.

In 1989, Seymour and I started something called "The Stonington Retreat." This was an intensive five-day summer learning retreat for teachers. We ran it for the next thirteen years. Seymour wanted to bring teachers to Deer Isle, Maine to reconnect with the joy of learning. Each session, we'd work with ten to thirty teachers. The weeks were amazing. We needed four or five days to allow us to get into deeper learning without the constraints we often faced: a limited time period in the school day, and the demand of everyday living. I discovered that giving my adult learners the change to find their own groove of learning was most effective. Just as I had already learned in teaching children, I was reminded that in our desire to impart knowledge, we often get in the way of learning. I started stepping back and listening more to what was really needed— in teaching both children and adults. Lessons became organized and a structure was found based on my students' interests rather than my own assumptions as to what was important to cover at a given time. Invariably, this method allows us to cover more material and do so more deeply.

I found combining my background in painting and sculpture with LOGO, LEGO Robotics, and other maker-y endeavors involving circuits, wires, and motors was really challenging and really fun for me. I initially had no curriculum, but fortunately as these disciplines were so unformed there were no curriculums to use. I had a lot of creativity and I loved coding. My students and I came up with projects that meshed coding with art; we designed New York City buildings, allowing the size of the buildings to be controlled using a concept called variables, and sharing each building with each class member. Every child could create their own New York City skyline using the work of each of their classmates.

In 1998 when I started Robofun®, I didn't quite realize that what I was trying to do was unusual. After all, I did it every day in my classroom and in the summer with adult students from around the world. I'm beginning to realize it was and it *is* unusual.

Robofun has taught **STEM to over 20,000 children** in our private studio on the Upper West side of Manhattan and in **over 300 schools across New York City**. I employ over 50 people who are as passionate as I am about empowering children. Our mission is to construct solutions that inspire students and teachers, and to address the widest possible range of academic standards through the use of leading-edge technologies. Robofun **provides children a teaching environment that satisfies their curiosity but also allows them to learn in a way that recognizes their individuality – the end result is their love for learning and a desire to repeat it over and over again**. We ignite kids' desire to explore, solve problems and learn. We invent, build, and have fun with cool technologies. Our philosophy of learning is rooted in constructivism, emphasizing a hands-on problem solving and project-based approach to learning that challenges students to think creatively, apply concepts, and actively "construct" meaning.

Our work has been funded by the National Science Foundation as well as the MacArthur foundation. We've collaborated with the MIT Media Lab, NYU's ITP, Columbia School of Engineering as well as the New

York City Department of Education. We've written curriculum for the Boys and Girls Clubs of America, and for the YWCA's Techgyrls program, and the original user's manual for Scratch, the language developed at MIT a few generations after LOGO.

In a world where children spend excessive amounts of time glued to digital devices and less time on interpersonal relationships, now more than ever we have to do our part to teach children to think creatively, problem solve, deal with frustration, collaborate and learn to design with new technologies. This is what I saw twenty-one years ago, and this continues to be Robofun's mission. We teach using robotics, coding, stop motion animation, Minecraft and circuitry, and this list will continue to expand and evolve. Our curricula have been used by thousands of children and we have trained countless teachers. These curricula and technologies are the platforms we use to get the ultimate result: kids who learn to love learning. We only hire teachers who love children. They must be present, connected, and excited by what they are teaching. They are able to take a simple project, such as making a robotic "Jack-in-the-Box" and see a moment to turn this into a class-wide Rube Goldberg sculpture. Our teachers (who we call mentors) are artists and engineers, and former camp counselors. They understand that although they are trained on our curriculums, they must also improvise, listen, and respond moment-by-moment to connect with each child. Above all, what unites them is their love for working with children and their love for learning.

What is most important to me, and is expressed through and by my team, is for kids to have a great experience. We know that each child comes with her own strengths and her need to feel that we recognize how she learns. We focus on teaching kindness by being kind. We focus on collaboration by working on group projects. We teach kids to have a growth mindset as we explore what it means to solve hard problems that can't be done in just two minutes. We shower kids with positive reinforcement, and we ensure that they always feel safe emotionally, intellectually, and physically.

When my own son was eleven, I was forced to pull him out of school, as I found my brilliant, dyslexic son's needs were not being met in school. I called myself the reluctant homeschooler. This began a deep dive into homeschooling as well as into making sure that every class we run a Robofun is done in a thoughtful, well-considered way that approaches learning from many different points of entry and supports many different types of learning. Flash forward to the present. My son, the homeschooler, just successfully finished his three semesters of college!

The other day, the doorbell rang, and a twelve-year-old came into the apartment having just walked our Golden Retriever. He proceeded to sit down and talk to me about why he loved Minecraft and how he thought we should incorporate more Minecraft classes into our programs. I agreed with him and invited him to consult with my staff as we developed the program. When he left, my son said, "Mom, you just keep collecting kids." He was right! I love what we do, and I am determined to continue my mission to ensure that we show children the joy of learning.

Laura Hart's Technology Tips

1. **Know your clients and think backwards for the best methods of providing them with what they need.** My clients are parents looking to provide their children with quality educational experiences that increase (or create) a love of learning STEM. In the creation of every course, curriculum and marketing campaign at Robofun, I make sure my team has that front and center in their mind.

2. **Look ahead and behind to find great tech applications for your children.** We live in a world flooded with new apps, new languages, new ways to use robotics with kids. Keeping up with this is very important, yet, not throwing away the baby with the bathwater is equally important. Do we get rid of Shakespeare because he is old? It is an important balance between NOT being seduced just because something is new and holding on to great learning materials.

3. **Network, Network, Network.** Fortunately, I really enjoy networking! It has made a huge difference for the growth and sustainability of Robofun. I have a Masters in Education from Harvard and have done consulting with the MIT Media Lab. Keeping those contacts along with all of my NYC clients and professional contacts is both interesting, to see how people's lives and careers evolve and helpful for the growth of Robofun.

4. **Find a way (or a place) to work where you are not interrupted and can think deeply about the challenges you are confronting.** Another female business owner once commented that running a business is a bit like breastfeeding 100 people at the same time. At times, it can feel like you are constantly pouring out ideas, directions, and not leaving anything left for yourself. That doesn't bode well for your ability to strategize and make good decisions. Having a quiet space you can go to work can increase your ability to make good decisions.

5. **Consensus is great at times, but trusting your gut and leading is equally important.** It is a fine balance between developing a team that feels trust and engagement and feels their ideas are listened to, and yet finding the moments to make unilateral decisions. Getting your staff to think like a CEO works for maybe 20% of your employees (or my employees) but in the end, it is your company, your profit and your debt that you need to manage. The balance is choosing those moments carefully so as not to erode the group motivation and group engagement.

6. **Be ready to pivot and embrace the process. When you need to pivot, you'll probably know that before the rest of your staff.** During Covid-19, we did a huge pivot; we took all of our classes in robotics, coding, animation and Minecraft and brought them online. Did my staff think we could do this in 2 or 3 weeks? No, but I lead and they made it happen. And it worked! We're now offering online classes for kids all over the world.

7. **Notice when you are really wrong and what you can learn from it.** Sometimes running a company feels like throwing spaghetti against the wall and seeing what sticks. I've made some colossally bad decisions and enough good ones so we are alive and doing well. It is easy to want to run away from the bad decisions. But that is the moment to dig in and learn: what went wrong? why didn't it work? What can we learn so we make fewer stumbles? Instead of looking at these with shame and embarrassment, it can be very interesting and enlightening and actually fun to take the lessons learned and apply it to the next set of decisions.

8. **Make sure you don't forget about recharging your own engine.** Running a company is hard work and very taxing. It is crucial that you value your own wellbeing and have methods to prevent your own burn out. Overworking is toxic yet addictive. I find I need to regularly schedule my days so I don't overwork. When I overwork I get less productive and less motivated.

9. **Join a group of other business owners and meet regularly.** I love talking to other business owners and sharing my challenges and successes. It is very energizing and it helps me boost my motivation and growth.

10. **Form a company wide book group.** We meet once a week for 30 minutes. We pick books in the field of Educational Technology as well as business books and novels that address education. It gets us all thinking together and raises our trust and knowledge of each

other.

11. **Treat your staff as you would like to be treated.** Employees who feel recognized and appreciated enjoy their work more and are more productive. I want my staff to know how much I appreciate their hard work. I've made a point during the pandemic to talk to each employee regularly to see how they are managing their mental health during this difficult time.

For my staff to be acknowledged and treated well sets an example for how to treat our clients, students and parents. It helps them feel pride in the work we're doing to help kids love to learn using creative technology applications.

About Laura Hart

Laura Hart has worked with teachers, students and technology for over thirty-five years. She lives in New York City and is celebrating the 22nd anniversary of founding her company called Vision Education & Media and commonly known as Robofun. Robofun helps children and teachers use technology creatively and effectively both online and in person.

Robofun has a studio on the Upper West Side of Manhattan where they run programs in robotics, coding, animation, circuitry and Minecraft. Robofun works with children after school, and on weekends and vacations. They also offer private lessons, small pod classes, and birthday parties. Additionally, they run over 130 programs a week in schools and community centers all over New York City. They offer in-school programs online or in person during the day and after school. They also deliver workshops to teachers and parents. Robofun has provided services to over 200 schools as a NYC Department of Education vendor.

Robofun has received two NSF grants and a MacArthur Grant. They collaborated with the MIT Media Lab on the development of the first scratch manual for coding and created the curriculum for the YMCA's Tech Girls program. They were the winners of the Educator's Choice Awards at the World Maker Faire of New York City for four years. In 2014 Sesame Street filmed an episode in our studio, and more recently, GMA and Thomson Reuters have aired stories based upon our work. In

2017 they were awarded a My Brother's Keeper Proposal in partnership with Makeosity and the NYC Department of Education to teach minority children and families about electrical engineering and robotics.

An early interest in coding led Laura to receive her Master's from the Harvard Graduate School of Education (1996) where she collaborated on projects with the MIT Media Lab. For over 30 years, Laura worked closely with the late Dr. Seymour Papert, MIT's creator of the Artificial Intelligence Lab and one of the world's foremost experts on technology and learning.

Laura has a Bachelor of Science in Studio Arts from Skidmore College (1983). Prior to founding Robofun, Laura directed the Computer Department at The Buckley School in Manhattan for fourteen years and was Director of the Technology Program at the Little Red School House and Elisabeth Irwin High School for two years.

Robofun's philosophy of learning is rooted in the model known as constructivism, which emphasizes a hands-on problem solving and project-based approach to learning. Because each of us experiences the world through a unique lens, the process of learning is not a uniform one. Learning focuses on central concepts rather than isolated facts, and high-level thinking is encouraged, once basic concepts are grasped. The teacher is the facilitator or the 'coach,' guiding students, offering feedback along the way, while encouraging students to analyze, interpret and make predictions about information they encounter. Laura's top priority is to ensure that each student's needs are met and that they leave Robofun as a more confident and eager learner.

Mandy Froehlich

"Technology is just a tool to support what we would like to do when we've had an innovative idea. Using technology is not necessarily innovative. Innovation is a way of thinking."

- M.andy Froehlich

Chapter 7

The Winding Road to Purpose

By Mandy Froehlich

I believe that there are only a lucky few that are blessed with knowing their purpose in life early on. Even if you know what you want to pursue as your *profession*, it doesn't necessarily mean you know what your *purpose* is. And once you find out, you're fortunate if those two go hand-in-hand. But purpose drives us forward with a passion that cannot be replicated by just being in a job. Purpose is what keeps us up at night and makes us anxious when we're not working within it.

I was an educator but I did not grow up teaching my stuffed animals and making up fake worksheets. I didn't dream of my own classroom. I worked in schools as a technology integrator and Director of Innovation and Technology, but when I was young my interest in computers went about as far as *The Oregon Trail*. If you would have told me I was going to work in schools I would have laughed at you and told you no, I was going to be a lawyer and I was going to go to Harvard. Back then, working with kids was nowhere on my radar.

But when your purpose is set and you start to veer off your path, the Universe has a way of gently nudging you back on. After having my own children and watching them grow and learn, I decided that I actually did want to teach them. So I went back to college and finished my degree with my four kids underfoot. Then a master's in Information and Communication Technology. Then another in Educational Leadership. I found both teaching and tech combined to be a strength of mine. It was a passion, but I had not yet found my purpose.

Currently, I am a consultant and work nationwide with districts in multiple capacities. Some of these are directly related to my experience in technology and leadership, but during my time in the education profession I went through the burnout and disengagement that so many educators experience. Yet, with a family, growing student loan debt, and the necessity of a steady income and insurance, I couldn't leave the profession. This realization begged the question: it's common to talk about educator attrition rates, but how many educators do we still have in the profession that want to leave and just *can't?*

At about the same time that disengagement was setting in, I started noticing some of the challenges that teachers just naturally go through in a day that would seem ludicrous in just about any other job situation— one of those being the verbal and physical abuse, even at the elementary level, that they were enduring on a daily basis. In some situations, students would become angry and scream or cuss at teachers and the student is removed, but the teacher was never given the emotional processing time to come down from the experience. Other situations involved students who relied so heavily on brutal physicality to communicate that the teachers and aides would have bruises all over their bodies and there was little to no additional support by the district to reduce this daily abuse. Also, I thought about the impact on the educators' mental health exacerbated by any potential underlying post-traumatic stress disorder symptoms that an educator might already be experiencing from past personal traumas. The potential negative consequences of adults who are stressed, demoralized, and burnt-out working with children that are in their care for most of their waking day can be catastrophic if not recognized and supported.

I knew that we could do better and we had to do more. I believe it's a fundamental human right to feel safe and be happy in what we do, and that definitely includes educators as they are not the martyrs that we so often make them out to be.

In my own experience, I had to defy the gravity of negativity and disengagement and make the choice to find ways to improve my mental health so I would have the ability to choose happiness every day. Because I fought back to loving my job without a handbook or support, I have created documented strategies that worked to re-engage the disengaged educator. I have been working for the last five years on supporting educator mental health, self-care and boundary setting, resilience, and desperately trying to destigmatize educator mental health issues. I've written several books that have a significant amount of research into educator mental health and the reasons why educators disengage. I have developed organizational structures for supporting educators in innovative practices, re-engagement, and resilience while continuing the demanding and excessively rewarding job of being an educator. And there, inextricably intertwined with teaching and technology, I found my purpose helping people reflect on their own mental health and if necessary, find the words and strategies they needed to start to heal. I learned the importance of connectedness, belonging, and balance, and how technology influences them. I discovered how mental health issues can impact innovative thinking and how if we are going to ask people to be innovative thinkers, we need to start in an area we may not have expected: supporting their mental health.

Many of the tips I provide fall into three categories: connectedness and belonging, innovation and technology, and mental health. I'd like to expand on my thinking in these areas.

Connectedness and belonging

We have unprecedented access to people around the world to learn and grow. While technology can feel so cold and unfeeling, I have grown a professional learning network and professional learning family that understands me and supports me. They make me stronger and better than I am by myself. Technology, for me, has been the conduit for me to connect with these people, and because of that, I credit it with giving me all the feels in spite of its cold nature.

Many times, I hear groans and grunts of disapproval when I suggest connecting on social media. The online bullying, the constant comparisons that happen when you see someone else succeed, the mindless scrolling can all feel like it adds up to evil, cult-like indoctrination. However, we have the control over how we use technology and social media. There is so much power in that control. First, we have the ability to be intentional with our scrolling, who we follow, and the posts that we allow ourselves to see. Second, we are the only ones that can control the way that social media makes us feel. We need to keep in perspective that when a colleague posts that they wrote a blog post that week, instead of feeling guilty that we didn't do the same, we can celebrate with that person and give them the credit they deserve for getting something done. Many times, we don't know the backstory and it's possible that was the only blog post they've gotten out for six months. Celebrate with them. Their successes do not make you less *unless you allow it.*

Innovation and technology

Our access to different technology tools have also given us the ability to create, iterate, and be innovative in different ways than we have created before and share our unique ideas with the world. I've seen artists draw, color, "paint" and form works of art utilizing a single stylus, touchscreen, and app. We not only have access to these new creative outlets but so do kids as well as the ability to share their creations with the world. They no longer need to tell their grandparents about the artwork, presentation, 3D model, or image they created. They just need to email it to them.

Of course, just using technology is not innovation. Innovation is a way of thinking. Technology is only the tool that can help ideas come to life. Without innovative thinkers, technology is just an expensive hunk of junk.

Mental health

Utilizing technology to support good mental health can seem like an oxymoron, but taking into consideration that we have control over how we utilize social media and the intention that we approach technology use with, we can absolutely use the tool to help us be more in-tune with our habits and self-care. There are, of course, apps that will teach better habits, help you self-manage, and do guided meditations. Also, utilizing the digital wellness elements of your smartphone can help guide you to bring more balance into your cell phone use. Finally, helping kids with cell phones learn these tools as well can ward off cell phone addiction and create more intentionality with how they use their technology as they mature.

My path wound me in so many directions that seemed unrelated at the time, but ended in a place where I had the opportunity to take what I had learned along the way and intertwine it with what, surprisingly, turned out to be my purpose. In the end, it had always been about helping people, anyway. I consider myself lucky that my passions and purpose were, like teaching, technology, and mental health, inextricably linked.

Mandy Froehlich's Technology Tips

1. As humans, connection and belonging is high priority. While technology can feel like it dehumanizes us, we have the opportunity to cultivate relationships in ways we haven't had before. Use that power and create those moments with intentionality.

2. Social media can help us connect, but it can also create feelings of negativity. We are the only ones who can control the way we allow social media to make us feel.

3. Learn how to follow and find people on social media who make you stronger and better. They will become your professional or personal learning community. Also, learn how to unfollow people who cause you stress and anxiety. Consider social media as one of the places where we have absolute control over the negativity that we take in.

4. We have the opportunity to create works in ways that the world has never seen before. Our creativity in using technology right now knows very few bounds. If we have the courage enough to share them, the whole world has the opportunity to enjoy our talents.

5. Utilize your phone as a tool to help you find balance, particularly if you seem glued to your phone. Seems counterintuitive? You have a tool that is willing to remind you to take a break. Understand your phone's digital health function and timers to help keep your usage in

check.

6. Technology is just a tool to support what we would like to do when we've had an innovative idea. Using technology is not necessarily innovative. Innovation is a way of thinking.

7. Use connection and information to help you grow as an individual. We have unprecedented access to information and people who can assist us in growing our personal selves whether that be through self-help or supporting our passions.

8. Try an online course through a university or online learning platform like Udemy. For a minimal cost, you can grow a business, learn a new skill, or enhance your current skill set.

9. Stop using the term "tech-savvy". It is a term that implies exclusivity for people who understand technology. In reality, we are all on our own journeys to learn if we are only willing to push a few buttons.

10. Teach kids to own their own digital wellness. They may view restricting their time as punishment because that's where their friends are. Instead, show them how to take responsibility for their time.

About Mandy Froehlich

Mandy Froehlich is a former educator, technology integrator, and Director of Innovation and Technology turned technology and mental health consultant. Her passion lies in reinvigorating and re-engaging teachers back into their profession as well as what's needed to support teachers in their pursuit of innovative and divergent thinking and teaching. Her work in mental health and how it impacts disengagement and innovative thinking and subsequently technology use is widely known in the education community.

She consults internationally with school districts and post-secondary institutions in the effective use of technology to support great teaching, technology department leadership and coaching, 1:1 rollouts and learning management system deployment, and the impact of mental health on innovation. She also consults with edtech companies on creating user-friendly and pedagogically sound technology for schools.

As a way to impact an upcoming generation of leaders, Froehlich is an adjunct for an organizational technology class for graduate students looking to become administrators. Her first book, *The Fire Within: Lessons from defeat that have ignited a passion for learning*, discusses mental health awareness for teachers. Her second book, *Divergent EDU: Challenging assumptions and limitations to create a culture of innovation*, is based on an organizational structure she developed to support teachers in innovative and divergent thinking. Her third book, based on educator engagement and mental health, is titled *Reignite the Flames: Finding our passion and purpose for learning among the embers* which has a recently released companion guide/workbook titled *The Educator's Matchbook: A weekly guide to reigniting your love of teaching, building resilience, and fighting burnout and disengagement.*

Kenyatta Powers-Rucker

Find a mentor to share their experiences, provide advice and help with advancement and your aspirations.

-Kenyatta Powers-Rucker

Chapter 8

My IT Journey

By Kenyatta Powers-Rucker

Although women make up nearly half (46.9 percent) of the total workforce they only make up 25 percent of the Information Technology (IT) industry. And this number is less than women working in IT in the 1980s. One of the several reasons that there is a shortage of women working in the IT industry is the lack of role models. It's hard to be what you can't see. I am sharing my journey and experiences of how I got into the IT industry to prove that it is not a "boys only" club.

I have been in the IT industry for over 20 years. Although I have been at the same organization for my 20+ years, I served in several different roles working my way up from a Jr Network Engineer to a Chief Information Officer (CIO).

My IT journey started when I took a typing class in high school. I liked it so much that I later took a computer programming class (although it wasn't much typing) which I found even more intriguing. This all led me to minor in Management Information Systems for the two years I attended college. Life happened and I couldn't return to college and I ended up working in an entirely different field. After several years at that job, I got laid off. Remembering how much I enjoyed my technology classes I decided to go back to my roots and attend a Microsoft boot camp. My time in Boot camp led me to get a few Microsoft certifications in server and workstation support. I then started applying for IT positions.

After many leads and interviews, I was led to a position on a government contract. I did not come close to qualifying for the position on paper because it required a degree and years of experience of which I had neither. I was so discouraged by all of the previous unsuccessful interviews I almost turned down this interview. I thought it was a complete waste of time mostly because I didn't check all of the boxes to qualify. But I decided to go and I walked into the room to an interview panel of three men. I immediately thought I was doomed. Surprisingly, the interview went pretty well but I still wasn't convinced I got the job. But a few days later I got the call. I was hired as a Jr. Network Engineer on a six month contract to support the Department's Y2K project.

This experience truly taught me not to underestimate myself. Although I did not qualify on paper, the panel saw something in me that they needed and decided to hire me anyway. This happens a lot. You do not have to be 100% qualified to be the best candidate for the position. There is always a human factor in the interview that you can display passion and enthusiasm and do not forget the power of soft skills.

A team of us was hired to install and update the Department's computers to ensure they were Y2K compliant. By no surprise, I was the only African American and the only female on the team. I was extremely intimidated by all of the men and the expertise I thought they had.

It turns out that we were all learning together. They were great guys and we learned a lot from each other. With that said, I still had to make sure I represented the African American community. I learned all that I could and I made sure I always professionally presented myself.

This was my first IT job, I learned something new every day and I loved it. I had no real experience but going through the boot camp classes and getting the certifications helped me a lot. It helped me to know and understand what I learned and not just memorizing it.

When the project was over and the six months were coming to an end the managers decided to extend my contract for another 6 months. I then started more desktop support to help support all of the new equipment we just installed and updated. By this time they hired a couple of more females and African Americans.

Two years and four contract extensions later, I was also building and maintaining servers and had received a couple of more certifications in Novell, the platform we were using at the time. The managers announced a permanent state position was opening up so I decided to apply for it. I got a job as a Computer Network Specialist as a state employee. I was mostly doing the same thing I was doing as a contractor.

After a year or so of being a specialist, two supervisor positions became available to supervise the Local Area Network unit that I was in. It never crossed my mind to apply for it until the manager suggested that I apply. She said because of my work ethic, I received certifications without anyone requiring it of me, and I wasn't in a clique per se, she felt I would be a great leader. I applied for the job and I got one of the supervisor positions.

We went through a management change and my direct manager was assigned to another department. I was now reporting directly to the new Director of the unit. We worked well together and he called on me for help in several different areas. Although I wasn't properly paid for it, I essentially became his Assistant Director.

I was a supervisor for a couple of years and the then CIO offered me and the Director new roles. My supervisor was offered the position of Deputy CIO and I was offered the position of Director. I was shocked, I had no idea this was coming but I was elated that he considered me for this position.

As Director, I am now overseeing the unit I previously supervised in addition to a couple of other units. I am now equal to or supervising my former peers and people that used to supervise me. It was a little awkward at first and there were a few people that challenged and tested me. So once again, I still felt like I had to prove myself. But I did my job and did it well. I had the support of management and that helped a lot.

I was a director for a couple of years and our CIO and Deputy CIO resigned and a female CIO was hired. The Department was growing so two more Deputy CIOs were also hired. After maybe a year, one of the Deputy CIOs left and I was hired as the new Deputy CIO.

As Deputy CIO, I now oversee six different units; Help Desk, Local Area Network, Wide Area Network, Telecommunications, Data, and Network Security, and Desktop Support. The other Deputy CIO managed the Application Development and IT Operations which included Project Management Office, Personnel, Budget, and Procurement. Although the other units were not under my direct supervision I had heavy involvement in most of them. I was attending the application development status meetings, managing my units budget, procurements, and personnel so I had a good understanding of most of it.

After a couple of years of being a Deputy CIO, the other Deputy CIO resigned and we did not hire anyone to replace her. So now I am the Deputy CIO for the entire Department. The CIO and I split the remaining units, she managed some of the IT Operations and she had heavy involvement in the Application Development. Because I was involved in these units all along it helped me tremendously when I had to directly manage them.

I was Deputy CIO for a total of around five years when the CIO resigned. I then became the CIO for the Department. Out of all of the advancements, this was by far the most challenging. I soon learned there was a lot the previous CIO was doing that I was not familiar with. As CIO, I was responsible for all aspects of technology in an enterprise environment for the Department. It included managing around 400 IT staff, supporting over 50 applications, for over 6000 users in over 100 locations across the state.

As I advanced in different positions throughout my career, I went through the proper hiring and interview process for each advancement. Also, other staff advanced right along with me moving into my previous positions. It was essentially advancements for everyone.

I have been CIO now for almost ten years. I have learned so much at DHS and my journey has allowed me to grow professionally. Although each position had its challenges, each put me in a better position to understand the needs of each unit as well as the department. This became very important as I transitioned into the different leadership positions.

Although I experienced female leadership in Technology we still were far from being the majority. I did not technically have a mentor through my career but I now see how important it is for women in technology, especially African American women to have mentors and role models. To help bridge the gap, I mentor and speak at different conferences to share my journey and experiences of my career to encourage and inspire other young ladies to also consider an IT career. It is important for women to see and hear from other women in the industry so that they can possibly see themselves doing something similar.

Kenyatta Powers-Rucker's Technology Tips

1. Always apply for the position. Even if you don't check all the boxes.

2. Certifications are a great way to learn IT knowledge and in many cases sought after more than a degree.

3. Highlight your soft skills as well as your hard skills. Soft skills are necessary to succeed in the workplace and are in high demand for many different types of jobs.

4. Interested in a certain position? Speak with someone already excelling in the position to really understand the skill set and the experience required as well as understand the day to day responsibilities.

5. Find your niche. Become extremely skilled in the area you enjoy and master it. Once you master that skill move on to the next. You will become the main known expert in that area in which your colleagues can trust.

6. Find a mentor to share their experiences, provide advice and help with advancement and your aspirations.

7. Learn from your mistakes. Never forget what went wrong so you can learn what to do right the next time.

8. Just entering IT? The Help desk is a great position to start. You will be introduced to all areas of technology and can get an idea of the

area that interests you the most.

9. Want to be promoted within your company? Identify a known problem and work to find a solution.

10. When interviewing, do your research on the company you are interviewing with. Incorporate what you've learned in the answers you provide. It's impressive to the panel that you've done your homework

About Kenyatta Powers-Rucker

Kenyatta Powers-Rucker is the first African American female to hold the position of Chief Information Officer (CIO) at the Maryland Department of Human Services (DHS). She functions as a senior-level technology executive combining project management, technical expertise, and business proficiency to develop, execute, and manage strategic, statewide technology projects for the Department.

Kenyatta has served DHS for over 20 years as both a consultant and an employee, providing technical expertise to the Department. As CIO, Kenyatta manages the Office of Technology for Human Services (OTHS), which supports the entire Enterprise IT environment for the Department that includes over 200 IT staff, supports around 40 applications, with over 7000 end users across 100 locations across the state.

She is also an advocate for bringing awareness to women and the underrepresented to the Technology Industry. She frequently speaks and mentors and aims to increase the number of women and minorities in the digital space by empowering them to become innovators in Information Technology and by providing the appropriate resources and opportunities.

In addition, Kenyatta found Taste of Technology, a program that is designed to expose and provide awareness to the underrepresented youth and women to the different careers, opportunities and role models in Information Technology. The Taste of Technology provides information on trainings, conferences, available programs, jobs and other opportunities to expand knowledge and expertise in Information Technology. The goal is to provide an environment to inspire the underrepresented youth and women to become innovators and technology professionals through targeted exposure programs and mentoring collaboratives.

Kenyatta is a best selling contributing author of "Women of Virtue Walking in Excellence" Volume II and "Step Into Leadership Greatness" and contributing author to "Women Who Lead in Technology" to be released April 2021. She is also a member of several Executive Information Technology (IT) Advisory Boards. She has also received many awards and accolades in addition to being featured by Wonder Women Tech Foundation as one of the "136 Black Innovators in STEM You Should Know and Support".

She is also a leader at her church and serves on several different ministries. Her love for God and her passion for empowering and supporting Community Development initiatives through ministry assignments has made her a staple in the community.

Patricia Brown

"Be an advocate for equity through Access and Opportunity."

- Patricia Brown

Chapter 9

Walking In My Purpose

By Patricia Brown

"**M**e a teacher, never in a million years!" Becoming a teacher was never a career that I envisioned for myself, in fact, it was the furthest thing from my mind. As a child I spent many years assisting my mother, who was an elementary school teacher, with her numerous teacher "duties." I graded papers, designed bulletin boards, tutored her students and performed any other task delegated to me. As a child, I absolutely loved it, as a teenager I grew to hate it and pledged that I would NEVER become a teacher. And now? 19 years later, I am on the greatest career path ever, dedicating my life to better the educational needs and environment of children. I AM a teacher! So, let me tell you how I got here.

I had a very traditional public education in Cincinnati, Ohio, and I graduated from one of the top 100 high schools in the country, but when I think back to my K-12 education, I don't remember having the opportunity to explore my passions. School for me was boring, but I was always compliant. Compliant, but not engaged. At age 11, my grandmother bought me my first computer, a Magnavox word processor. This was an anomaly, because people were not buying kids computers back in the 80's and 90's. But my Grandmother Blue invested in me, and that investment helped me tap into a passion that I don't think I would have realized at such a young age. She created the space, and gave me an opportunity, that helped me find my sweet spot; where my passion, the thing I loved, and my purpose all came together. That Magnovox word processor became my world, I created flyers, and brochures, and by the time I was 14 years old I had started my own design business.

Upon graduation from high school, I knew I wanted to attend college, but was unsure as to which subject area(s) I wanted to focus my studies on. At the end of my first semester at Tennessee State University, I declared my major in Business Information Systems and planned to work in the field of technology. However, during college, my work experiences always seemed to involve youth. I remained active with my home church and during summer vacations I returned to my youth group where I served as a mentor. As a mentor, I supervised teens ages 12-17 in an academic enrichment and career planning program and provided hands on instruction with computers and other technology. At the beginning of my junior year of college, a new degree program was developed which offered students the opportunity to focus on the business aspect of technology, as well as become certified to teach Business Education. I took advantage of this new opportunity, thus declaring a dual major in

Business Information Systems and Secondary Educational Technology. I instantly became intrigued with discovering how students learned and the many ways technology enhances creativity. The year 2001 was a whirlwind for me. Upon graduation, I got married, relocated to St. Louis, Missouri, had my first son, and began teaching at a school for academically gifted students in the St. Louis Public Schools. Initially, I was extremely nervous and intimidated by the thought of teaching "gifted" students. However, I learned that most of my students entered our program with extensive computer knowledge, but alternatively struggled in substantive academic areas. I was certain that technology could be used to support their learning. I wanted to provide children with accurate, clear and concise instruction, so they could see how technology is applied in their daily lives. My goal was to provide them with opportunities to visualize and conceptualize the impact technology has on our society.

During that time I allowed others to minimize what I did as a profession, as if teaching technology is not as important as other subjects. However, I quickly learned that teaching new technologies can provide meaningful learning experiences for all children, especially those at risk. Technology can help students to develop higher order skills and to function effectively in the world beyond the classroom. Using technology for meaningful activities also helps to integrate a variety of disciplines, more closely resembling activities that people undertake in the world beyond the classroom. I had so many ideas. I would attend conferences and workshops and learn about the latest and greatest, and would return back to my school to a zero technology budget, and little support. I knew that if I wanted to teach kids how to use technology in creative ways they had to have access to it. That access shouldn't be determined by what their zip code or what their parents did. After a few years of frustration, I made a decision. I decided that I would do whatever it takes to provide opportunity and access to my students. In 2009, I wrote and received a technology grant for over 25,000 for my classroom. I used that grant to purchase technology like iMac computers, and flip cameras. I wanted my students to have the same access to technology as the students in the affluent districts in St. Louis, otherwise known as the "lucky zip codes." One day I was having a conversation with one of my colleagues, Katrina Love, an amazing 7th grade Algebra teacher who sadly passed away a few years ago. She was explaining to me how she created a dance for her students to teach them the direction of a slope, because they were having such difficulty with that concept. I asked her wouldn't it be cool if our students made a music video? Not only would it help our current students, but you would have a resource for your future students to make a connection with. That conversation sparked an idea. Our students wrote and produced a music video! This was a game changer. Not only were they learning the content, they were making a connection, and gaining lifelong skills. I believe all children are born with the capacity to be creative, but their creativity won't necessarily develop on its own. It needs to be nurtured, encouraged, supported. It's like how a gardener takes care of his plants, by creating an environment in which the plants will flourish. I believe when you are giving opportunities for students to learn about their interests and explore their passions, and connect it to content, that's where creativity can flourish. During my nine years as a classroom teacher, I continued to learn and grow. My vision was that technology would no longer be viewed as a separate entity, instead it would become integrated into all facets of education.

During that time, I had four more children, earned two graduate degrees, and I became a local conference presenter. Needless to say it was challenging. There were not many presenters that looked like me. Let alone the audience. I spent most of those years in isolation, sometimes feeling as if I was on an island by myself. So how did I get through it all? God's grace. I had a purpose. One of my favorite scriptures is Jeremiah 29:11 "'For I Know The Plans I Have For You' Declares the Lord, 'Plans to Prosper You and Not to Harm You, Plans to Give You Hope and a Future. I kept this scripture close to my heart, and leaned on it when times were tough.

In 2010, an amazing opportunity fell into my lap, and I left the classroom for my dream job as a Technology Integration Coach, in an affluent district in St. Louis. I have the best job in the world, I get to help teachers implement technologies in the ways I always envisioned; as a tool to engage digital natives, and create authentic real-world learning experiences. Through the years I have obtained many awards, accolades, and extraordinary learning opportunities. I continued to perfect my craft and stay on the cutting edge of technology which led to various opportunities, speaking engagements, and business partnerships. But there was something missing. Connections. I spent the majority of my career providing those connections for others, but who do I get to connect with? Where are the people that look like me, who share the same vision, values, and beliefs about teaching and learning with technology? Do they even exist out here in cyberspace? I soon got my answer, and this connection changed my life. In 2016, I was nominated for an award honoring my work in Digital Media, and as I was looking on the website I came across my now good friend Sarah. After a DM on Twitter, she introduced me to a group of other "Edtechs" Rafranz , Carla , Regina S, and Nicol and who were all black women in edtech connecting on this cool app called Voxer. We called ourselves The United Edtech. We created a safe space to share ideas, celebrate each other, vent, and even have healthy debates. These women became my family, and I learned so much from them. Through the years, we became very active in the International Society of Technology in Education (ISTE) organization, and joined the leadership team of the ISTE Digital Equity Professional Learning Network. Pretty soon this space grew, as we invited more educators of color, and began to expand our reach, as we advocated for equity. I finally felt the belonging I so much desired early on in my career, and I am eternally grateful for these divine connections.

When I watched the movie "Hidden Figures," I felt empowered because it shed a new light in history, and gave a powerful message of the importance of highlighting and celebrating diverse voices and perspectives. It also allowed me, a black woman, to see myself as a scientist, mathematician, and a creator. It also erased the negative stereotype that women, specifically women of color did not excel in math and science, or have a place in STEM or history. This movie sparked my passion to create a space, and opportunity, for students to learn about STEM and career opportunities by highlighting "hidden figures" and connecting students to African American STEM professionals. Because of this, I created "STEM Gems" , an after school program for elementary girls of color. This was no easy beginning, as there was pushback from others who didn't understand why I was being intentional in providing this opportunity for girls of color. It's about equity. I realized that it is very hard to be what you can not see.

As black woman, and a mom of five boys, it is extremely important for me to create innovative experiences for all students, but especially for students of color. I am fortunate to have experience in school districts from both ends of the financial spectrum. As an educator in an urban district with a limited technology budget, I had to remove a huge financial barrier by seeking technology resources from private agencies the budget restraints kept out of reach. This inspired a few teachers to use technology to differentiate and enhance student learning. In my current affluent district, where the technology budget is plentiful, I quickly learned that purchasing the most cutting edge technologies does not guarantee that it would be used effectively. Both experiences have enlightened me on what essentially removes these barriers. My greatest contribution to education is my commitment to my students. With several advanced degrees in Technology I could easily be working in corporate America making twice the money, but instead I am proud to say that I am an educator. Why did I become a teacher? I wanted to make a difference. As a black woman, I have the opportunity to influence an often labeled "troubled" generation. I am very dedicated to closing the generational gap, and serving as a positive and professional role model to all young people. Now then ever before it's important to become an advocate for children, and create equitable learning environments. Through my educational experiences I had the pleasure to learn under very instrumental teachers who inspired and supported me. Like my mentors, I am confident that I will continue to play a pivotal role in the lives of others inside and outside of the classroom.

Patricia Brown's Technology Tips

1. **Reflecting on your own biases and assumptions**. Recognize that everyone holds some types of bias based on their personal background and experiences. It starts with you as an individual and understanding that you need to decenter yourself --ego aside...and reflect. Ask yourself, how does my own social location (race, class, gender, religion) shape my mindset about teaching and learning, the students or districts you are serving, and the practices you act out? Consider how a stereotypical image of a scientist could unintentionally create biases and assumptions about students.

2. **Be an advocate for equity through Access and Opportunity.** Digital equity can be described as Providing Access and Opportunity for devices, and ensuring everyone has knowledge of the possibilities through technology. But it's not just about handing a kid a laptop, it's about understanding the systems that have led to the lack of access, and advocating for change.

3. **Be culturally relevant and responsive**. Empowering your students to use their knowledge of their selves, culture, identity and personal interests to make connections with what is being taught. Forming relationships with students is important to establish a culture for learning. Every one of your learners should feel a sense of belonging and inclusion in every lesson you design.

4. **Foster an environment to create.** Creativity needs to be nurtured, encouraged, and supported. It's kind of like how a gardener takes care of his plants by creating an environment in which the plants will flourish. Similarly, you can create a learning environment when you are giving opportunities for students to learn about their interests, and explore their passions, creativity can flourish.

5. **Let it Flow.** Sometimes that creativity flows in unexpected times. Create opportunities to show what they know in creative ways. Allow for student voice and choice.

6. **Create a Positive Classroom Community-** When you are designing lessons, think about how we are using technology to ensure that all of our students' learning experiences are meeting the social and emotional needs of students who have been abruptly put into learning isolation since March.

7. **Provide the Space and Opportunity.** Are you creating opportunities for learners to engage in designing & making, and not just listening and observing? Are you offering opportunities for learners to engage in interactions with others as an audience? To effectively prepare our students for the future, and use technology productively, we have to move our students from passive to active users of technology.

8. **Be Intentional.** Create the space and opportunity for girls to thrive!

9. **Expose students to the possibilities through Stem.** There are a

million stem jobs that go unfulfilled every year. If we want to get kids interested in STEM jobs, and really increase the number of students going into STEM careers, we have to introduce them to the possibilities.

10. **Be inspirational.** The world is evolving more and the most complex technologies are now available at our fingertips. Our kids are witnessing the worst technology that they will see in their lifetime, share stories and experiences of the power of technology.

Remember this quote by Maya Angelou. You can't use up creativity. The more you use, the more you have!

About Patricia Brown

Patricia Brown, known as "MsEdtechie is an equity-minded educator. Patricia is a national keynote speaker, and the founder of MsEdtechie Consultants LLC. Patricia serves as a current member of the International Society of Technology in Education Board of Directors, and the Smithsonian STEM Advisory Committee. She was selected as a National School Board Association's 20 to Watch in EdTech, an Apple Distinguished Educator, and a Google for Education Certified Innovator & Trainer.

Dr. Sonja Ann Jones

You never know how close you are to making that dream happen.
Sometimes it is right around the corner. Tenacity is key!
~ Dr. Sonja Ann Jones

Chapter 10

Land on Your Feet and Progress Forward

By Dr. Sonja Ann Jones

Many women find that the journey in the technology industry can be filled with successes, challenges, barriers and obstacles. I can definitely relate with my journey in technology. My journey in technology started when I was a little girl. I always had an aptitude for mathematics and the sciences. Math was my favorite subject in school. In fact, in elementary school I was already taking calculus.

I continued to pursue my coursework in math and the sciences and I started to take college classes in high school. I quickly decided that I would work toward getting a BS in Mathematics at the University of Minnesota. While doing so, I started to learn Statistics. It was the first time as a student that I struggled with a course. I reached out to my professor for assistance at the time and he told me I would never pass his course because I was a woman and that I didn't have the mind for statistics. I was so infuriated I took the class again with another professor and earned an A. I then decided to get a BS in Statistics as well.

Not only did I get both degrees but I was also the only one chosen for the Paul Cartwright Award in my graduating class of 1000 college students. The award is for the most well-rounded student for success in coursework and also for contributions to the community.

I co-founded the first national sorority in America for women in technology called Alpha Sigma Kappa – Women in Technology at the University of Minnesota. I worked with a few of my friends who were also women in technology to truly make a difference for Women in STEM. We formed an organization that supports women both personally and professionally. The organization has been so successful that there are now 14 chapters in the USA and it has touched thousands of women's lives since its inception on May 1, 1989. When I received the Paul Cartwright award the Statistics professor that told me I would never make it because I was a woman, was on the same stage when I received my award and both my degrees. I did not back down and proved to myself and everyone else that women do have a place in technology. I never gave up and stayed dedicated to my goals to succeed. I had faith in myself that I could do it and make a difference for other women who were also facing the same challenges in technology.

After college, I struggled, like many, on figuring out what career would be best for me. I started as an actuary and then a computer programmer and found out that for me they both weren't a good fit. They weren't a good fit because I really loved to work more with people than behind the computer. A few years later, I was very fortunate to land into sales engineering, where I was able to work with my sales, business, and technology skills. I loved helping customers solve their business problems with analytical software. I dove into the business intelligence and the data warehousing world. I also eventually got into big data and cybersecurity.

I have had a good career over the last 20+ years in sales engineering, sales and sales management but I still have faced many challenges and obstacles in the technology world. Being a woman in technology, one thing you have to be aware of is you are a minority and unfortunately gender biases still exist today. In order to survive, I have learned many things about myself and the business.

I have continued to keep learning to make myself competitive. This is essential because I quickly learned a woman in technology always has to continue to prove themselves. I then earned a MBA and also a Doctorate in business where my dissertation focused on Women in Technology. I learned a lot through my dissertation and I am now writing my own book to release the findings of my research. In addition, I continue to learn and keep my technology and business skills fresh with many books and outside courses.

Real world experiences have really taught me a lot in both my personal and professional life. I have been faced with so many challenges, gender bias and unfair treatment but I have never given up. In fact, when the going has got tough, I have rebranded myself and adjusted my goals accordingly. I still have always landed on my feet and I have progressed forward.

I am a firm believer that no matter what happens in this crazy world of technology that I am true to myself and I am a firm believer of always giving back. It is so important to give back, assist and make it easier for other women that are in technology or want to get into technology. I have mentored many students and career women to get to the next level in their technology careers.

I even created my own organization called "Nonstop 4 the Top" dedicated towards these efforts. I help women in technology, business and entertainment to get to the next level. I love to mentor, sponsor and truly make a difference in other women's lives and careers.

In addition, I have been very actively competing in pageants throughout my career to promote my platform which is "Women in Technology". I really took that platform to the next level when I won Mrs. Corporate America in 2009. I spent my year long reign dedicated to advocating for women in technology. I focused on getting the word out there in the media and in the technology corporations. I was featured in many magazines, newspapers, podcasts, radio and television shows. To this day, I still continue to mentor and help women navigate this career. I continue to appear in the media, speak to organizations, contribute to articles and books and promote women in technology.

For any woman who wants to succeed in technology, you have to have thick skin and never give up when faced with the obstacles, challenges and biases.

In recent years, there has been some improvements but there is still a lot of work that needs to be done to give women a fair shake in technology, leadership and executive board roles.

If you are a woman interested in technology make sure you are ready for a crazy ride and never give up on your dreams and goals. You can do it if you really want it and want to put your mind to it. Trust me, I know from experience.

Dr. Sonja Ann Jones' Technology Tips

1. **Be confident in yourself.** This is key when others don't believe in you. If you don't believe in yourself, no one else will. Also toot your own horn when you have successes! Sometimes women fail to do this and it is a must! Take credit for your successes!

2. **You have a rightful place, you earned it.** You worked hard to get where you are at and you deserve to be there like anyone else. Don't let anyone else tell you otherwise!

3. **Never give up!** You never know how close you are to making that dream happen. Sometimes it is right around the corner. Tenacity is key!

4. **Never stop learning.** Continue to learn via books, online training, in person classes and from your peers. In order to stay competitive, you have to stay current!

5. **Network as much as you can.** Always keep up your network via social media, phone calls, email and getting together for impromptu meetings. A good rule of thumb is never eat alone. This keeps you networking.

6. **Always give back and help others.** Make sure to give back and make a difference. This not only helps others but can also help you leave a legacy!

7. **Keep your technical and soft skills sharp** – Again make sure you keep up to date on as many skills as possible. The combination of technology and business skills is gold!

8. **Always be open to new opportunities** – You never know when the right opportunity comes around, so you have to be ready to take advantage of it!

9. **Always be aware of things going on around you** – Always be mindful of what is happening in your workplace and industry. Stay informed because this helps you do your job well and also leads to other opportunities.

10. **Find the right balance** – Remember we are all human and we need to keep a good life balance to be productive. Enjoy life. Work and play hard!

About Dr. Sonja Ann Jones

Dr. Sonja Ann Jones is multi-talented and has over 25 years of experience in sales, business, and technology. She has advanced many companies towards their mission and goals via her consultative sales experience and has been essential in helping companies meeting their business requirements with the grown and evolution of technology solutions. She has expertise in Security, Big Data, Data Warehousing, and CRM. She has worked for many prominent companies.

http://www.linkedin.com/in/sonjaannjones

Dr. Jones recently completed her doctorate at Argosy University in Business focused on entrepreneurship and her dissertation was published with her groundbreaking studies around Women in Information Technology Leadership. She also has been published in five different books which include "How to Break a Glass Ceiling Without a Hammer:, "How to Survive When your Ship is Sinking", "Successful Women in Business", "Corporate Queens, What Business School Won't Teach You: and "Uncuffed – Surge of Power. Currently she is working on publishing her own book. Dr. Jones also has an MBA in entrepreneurship from the University of St. Thomas and two Bachelor's degrees in Mathematics and Statistics from the University of Minnesota.

Being a huge advocate for women in technology, she is the co-founder of the first technical sorority in America – "Alpha Sigma Kappa –

Women in Technical Studies" which has touched over 1,000 women's lives. She also recently created her own foundation, called "Nonstop 4 the Top". Her organization is dedicated to helping women succeed in business, entertainment and technology. Dr. Jones also mentors' women through Everwise and her college alumni programs.

In addition, Dr. Jones was crowned Mrs. Corporate America during her reign. She was a speaker and advocate for her women's platform through many forms of media that included television, internet, books, magazines, and many internet publications. Dr. Jones has also been instrumental on the diversity board of SAG-AFTRA helping women to be successful in entertainment as actresses, directors, and producers. Dr. Jones also has had an extensive background in the entertainment field as an actress and producer.

In her spare time, Sonja loves to act and is very competitive running marathons and playing tennis.

She also loves to work on her roses in her garden, travel and she enjoys time with her husband, family, and friends.

Dr. Sonja Jones' goal is to make a true difference in the world and have a positive impact on others. Sonja firmly believes if you work hard that dreams can come true.

@imsonjaannjones – Social media www.sonjaannjones.com
IMDB - https://www.imdb.me/sonjaannjones
https://www.linkedin.com/in/sonjaannjones

AFTERWORD

To say that this decade has started with a bang would be the understatement of the 21st century. For far too long, technology has been treated by many as an add-on, or "one more thing," but the pandemic of 2020 has brought to light just how important technology is in today's world. It is a utility, and digital equity has gained prominence as a topic of conversation. The past year has shone a light on both glaring inequities *and* pockets of innovation. In response, people around the world have increasingly banded together to help one another by connecting and readily sharing information.

This is where my sisters in technology have truly stepped up to the plate, as you have seen in the previous chapters. Hopefully, as a reader, you have gleaned insight into their journeys, as well as helpful takeaways that will be useful in your own practice. Connect with them, as well as with others who inspire you, and let's all continue to learn and grow together!

Dr. Sarah Thomas

Dr. Sarah Thomas

If you use social media, there is most likely a network of educators already there connecting and sharing best practices, as well as supporting each other and learning and growing together. Find these groups in spaces you already feel comfortable, then expand outwards as you learn more.

- S. Thomas

ABOUT DR. SARAH THOMAS

Sarah Thomas, PhD is a Regional Technology Coordinator in a large school district in Maryland, and the founder of EduMatch, a project that empowers educators to make global connections across common areas of interest. She has spoken and presented internationally, participated in the Technical Working Group to refresh the 2017 ISTE Standards for Educators, and is a recipient of the 2017 ISTE Making IT Happen award. She is also an Affiliate Professor at Loyola University in Maryland. Sarah is a co-author of the ISTE Digital equity series, *Closing the Gap*.

ABOUT THE VISIONARY

Dr. Sharon H. Porter

DR. SHARON H. PORTER

Dr. Sharon H. Porter (Dr. Sharon), educator, author, publisher, and host, is the President of SHP Enterprise, the umbrella entity of Perfect Time SHP LLC, Coaching, Consulting, and Book Publishing Form and SHP Media and Broadcasting. She is the Executive Director and Founder of The Next In Line to Lead Aspiring Principal Leadership Academy (APLA), where she trains, mentors, and coaches assistant principals from across the U.S. who desire to take the helm as principal.

Dr. Sharon is Co-Founder, owner, and Editor-In-Chief of Vision & Purpose LifeStyle Magazine, is the host of The I Am Dr. Sharon Show, and is Cofounder and Vice-President of Media and Communications for WNM Ventures LLC.

She has over 30 years of experience as a school principal, Leadership Development Coach, assistant principal, Instructional Specialist, Curriculum Coordinator, and elementary and middle school classroom teacher, and is currently serving as an elementary principal in Maryland. She is the author, visionary, and publisher of The Next In Line to Lead book series, The Women Who Lead Book series, and The HBCU Experience Anthology, book series.

Dr. Sharon is a graduate of Howard University, Walden University, Johns Hopkins University, National-Louis University, and Winston-Salem

State University. She is a part of the 2019 Harvard University School of Education Women in Leadership Cohort.

She is a member of the Forbes Coaches Council, International Association of Women (IAW), American Business Women's Association (ABWA), Professional Women of Winston-Salem (PWWS), and Delta Sigma Theta Sorority, Inc. She serves on the Board of Advisors for The Women of Prince George's and Envision, Lead, Grow (ELG), Inc, and is a part of the Executive Team of Black Women Education Leaders (BWEL), Inc.

www.ingramcontent.com/pod-product-compliance
Lightning Source LLC
Chambersburg PA
CBHW032330210326
41518CB00041B/2057